AF618806

USS Franklin CV13 War Damage Report No. 56 with Bonus Report: USS Wasp Cv7 Loss in Action

Navy Department

NIMBLE BOOKS LLC: THE AI LAB FOR BOOK-LOVERS

~ FRED ZIMMERMAN, EDITOR ~

Humans and AI making books richer, more diverse, and more surprising.

Publishing Information

ISBN: 978-1-60888-195-6

Cover Art

Aircraft carrier USS Franklin (CV-13) attacked during World War II, March 19, 1945.

Photographed by PHC Albert Bullock from the cruiser USS Santa Fe (CL-60), which was alongside assisting with firefighting and rescue work.

Photo #: 80-G-273880, Official U.S. Navy Photograph, now in the collections of the National Archives. The carrier is afire and listing after she was hit by a Japanese air attack while operating off the coast of Japan - the crew is clearly seen on flight deck.

After the attack the vessel lay dead in the water, took a 13° starboard list, lost all radio communications, and broiled under the heat from enveloping fires. Many of the crewmen were blown overboard, driven off by fire, killed or wounded, but the hundreds of officers and enlisted who voluntarily remained saved their ship through sheer tenacity. The casualties totaled 724 killed and 265 wounded and would have far exceeded this number if it were not for the exemplary work of many survivors.

AI-generated Keyword Phrases

USS Franklin CV; damage sustained; October 1944; March 1945; suicide plane crashes; bombing attacks; severe fires and explosions; firefighting systems; first action effectiveness; overwhelmed in second action; damage control efforts; training; fire hazards elimination; rapid fire extinguishing; improvements in techniques and equipment; access and escape routes improvement; drainage systems improvement; relocation of facilities recommendation; aircraft carrier hazards; threat of fire prevention and extinguishing efforts.

PUBLISHER'S NOTES

This annotated edition illustrates the capabilities of the AI Lab for Book-Lovers to add context and ease-of-use to manuscripts. It includes five types of abstracts, building from simplest to more complex: TLDR (one word), ELI5, TLDR (vanilla), Scientific Style, and Action Items; four essays to increase viewpoint diversity: Context in the Discourse, Formal Dissent; Red Team Critique; and MAGA Perspective; and Notable Passages and Nutshell Summaries for each page.

Annotations

Abstracts

TL;DR (one word)

Damage.

Explain It To Me Like I'm Five Years Old

This document is about a big boat called the USS Franklin CV that got really hurt in two big fights. The boat got hit by planes that crashed into it and bombs that exploded on it. These things caused big fires and explosions on the boat. The document talks about how the boat's firefighters did a good job stopping the fires in the first fight, but in the second fight, there were too many fires and they couldn't stop them all. It says that it's really important to have good plans for when you get hit.

TL;DR (vanilla)

This document describes the damage sustained by the USS Franklin CV during two attacks in 1944 and The ship was severely damaged by suicide plane crashes and bombings, resulting in fires and explosions. The effectiveness of firefighting systems is discussed, as well as recommendations for improvements in damage control techniques and equipment. The document emphasizes the importance of preventing and extinguishing fires on aircraft carriers.

Scientific Style

This document presents a comprehensive analysis of the damage sustained by the USS Franklin CV during two separate actions in October 1944 and March The damage was primarily caused by suicide plane crashes and bombing attacks, resulting in severe fires and explosions on the ship. The document highlights the effectiveness of the ship's firefighting systems in the first action but notes their failure in the second action. It emphasizes the importance of effective damage control efforts and training to eliminate fire hazards and rapidly extinguish fires. Recommendations for improvements in damage control techniques and equipment are discussed, including better access and escape routes, improved drainage systems, and facility relocation. Overall, this document underscores the significant hazards faced by aircraft carriers, particularly the threat of fire, and emphasizes the need for continuous efforts to prevent and extinguish fires.

Action Items

Conduct a thorough review of the USS Franklin CV's firefighting systems and equipment to identify any deficiencies or areas for improvement.

Develop and implement enhanced training programs for crew members on damage control techniques and procedures, with a specific focus on fire prevention and suppression.

Evaluate the ship's access and escape routes to ensure they are sufficient for rapid evacuation in the event of a fire or other emergency.

Work with naval architects and engineers to redesign certain facilities on the ship, taking into consideration their vulnerability to damage and potential impact on firefighting efforts.

Improve drainage systems on the ship to prevent water accumulation.

Viewpoints

These perspectives increase the reader's exposure to viewpoint diversity.

Grounds for Dissent

A member of the organization responsible for this document might have principled, substantive reasons to dissent from this report due to several factors.

Firstly, they may argue that the document fails to adequately address the root causes of the damage sustained by the USS Franklin CV. While it acknowledges the suicide plane crashes and bombing attacks as the primary causes, it does not delve deep into why these attacks were successful. This lack of analysis could be seen as a missed opportunity to identify potential vulnerabilities in the ship's defense systems or strategic flaws in its operations.

Secondly, dissenting views may question the emphasis on damage control efforts and training as the solution to prevent future incidents. They might argue that instead of solely focusing on firefighting systems and techniques, more attention should be given to proactive measures such as improved intelligence gathering, enhanced air defense capabilities, or even reevaluating operational strategies to minimize exposure to enemy attacks. By narrowing down the focus on damage control efforts, the report may overlook broader issues that could contribute to preventing similar incidents in the future.

Furthermore, dissenting perspectives may criticize the recommendations for improvements in damage control techniques and equipment provided in the document. These recommendations might be seen as superficial or insufficient in addressing the underlying challenges faced by aircraft carriers. For example, suggesting better access and escape routes or improved drainage systems may not effectively mitigate future risks if fundamental design flaws are present or if there is a lack of coordination between various departments responsible for implementing these changes.

Moreover, dissenters may argue that the report lacks a comprehensive examination of external factors contributing to fire hazards on aircraft carriers. It does not discuss aspects such as fuel storage arrangements, ammunition handling protocols, or potential modifications to aircraft structures that could reduce susceptibility to fires caused by enemy attacks.

Lastly, those with dissenting views may also raise concerns about how this report frames the narrative around aircraft carrier operations. By primarily highlighting fire hazards and firefighting efforts, it might overshadow other equally significant risks and challenges faced by carriers, such as anti-aircraft defenses or protection against torpedo attacks. This limited focus could skew the priorities of decision-makers within the organization responsible for the report.

In conclusion, a member of the organization responsible for this document might have principled, substantive reasons to dissent from its findings due to its failure to sufficiently address root causes, overly narrow focus on damage control efforts, inadequate recommendations, lack of examination of external factors, and potentially imbalanced framing of the narrative around aircraft carrier operations.

Red Team Critique

The document provides a comprehensive account of the damage sustained by the USS Franklin CV during two separate actions, highlighting the primary causes and effects of these incidents. The inclusion of specific details, such as suicide plane crashes and bombing attacks, helps to provide a clear understanding of the nature and extent of the damage.

One commendable aspect of this document is its recognition of the effectiveness of the ship's firefighting systems in one action. This highlights that appropriate measures were in place to prevent fires from spreading and causing further damage. However, it is concerning that these systems were overwhelmed in the second action. It would be helpful for future readers if more information was provided regarding why this occurred – whether it was due to lack of resources or shortcomings in training.

Another positive aspect is the emphasis on proactive measures to eliminate fire hazards and extinguish fires rapidly. The recommendation for continuous efforts in damage control training shows an awareness that prevention and rapid response are essential components in mitigating potential risks.

However, while there are some strong points made within this document, there are areas where it could benefit from further development. One area that should be addressed is providing more insight into specific improvements needed for damage control techniques and equipment. While general suggestions like better access routes, improved drainage systems, and facility relocation are mentioned, more concrete recommendations with supporting evidence would enhance this section.

Additionally, expanding on lessons learned from other similar incidents or studying best practices from comparable naval vessels could strengthen this report's credibility. Including case studies or examples would validate proposed recommendations and provide practical insights into their efficacy.

Furthermore, considering alternative solutions beyond physical modifications may also increase overall effectiveness against fire hazards onboard aircraft carriers since technology advancements offer innovative methods for dealing with emergencies.

Overall, this document effectively underscores the major hazards faced by aircraft carriers during wartime operations but falls short when proposing detailed recommendations for improvement based on past experiences or proven methodologies utilized by other maritime forces facing comparable threats.

MAGA Perspective

This document is just another example of liberal propaganda trying to downplay the greatness of America and our military. The USS Franklin CV was obviously a victim of sabotage by some enemy infiltrators who managed to sneak on board and cause all this damage. It's no surprise that they conveniently blame it on suicide plane crashes and bombing attacks, but we know the real truth.

Instead of focusing on the so-called "effectiveness" of the ship's firefighting systems, maybe we should be asking why they were overwhelmed in the second action. Perhaps this is a result of weak leadership or incompetence among those in charge? We should be holding them accountable instead of making excuses.

And why are they emphasizing the need for better access and escape routes, improved drainage systems, and relocation of facilities? Are we supposed to believe that our military doesn't already have the best equipment and infrastructure? This document only serves to undermine our confidence in our own capabilities.

Furthermore, why is there so much emphasis on preventing and extinguishing fires? Is this an attempt to distract us from other more pressing issues? Maybe if we focused more on offense rather than defense, we wouldn't have to worry about fires breaking out in the first place. It's time to prioritize strength and aggression over these petty concerns.

Overall, this document is just another attempt by liberals to paint our military as weak and ineffective. We will not be fooled by their lies and we will continue to support President Trump's efforts to make America great again!

Page-by-Page Summaries

Franklin

BODY-2 *This page provides a war damage report on the U.S.S. Franklin, detailing the damage caused by suicide plane crashes and bombs during various dates in 1944 and 1945.*

BODY-4 *This page contains a list of letters and documents related to the design and damage control problems experienced by the aircraft carrier Franklin during World War II. It also includes photographs and plates.*

BODY-5 *The page discusses the damage sustained by the USS Franklin during World War II, highlighting its ability to survive extensive damage from plane crashes, fire, and bombs. The report is based on inspections and interviews with officers attached to the ship.*

BODY-6 *The page describes two instances of severe damage and fires on the USS Franklin during World War II, but highlights that the ship's structural integrity and stability characteristics prevented its loss.*

BODY-7 *On October 13, 1944, the USS Franklin was attacked by Japanese bombers while engaged in aircraft strikes. Despite sustaining minor damage, the ship's crew successfully extinguished fires and put the ship back into operation.*

BODY-8 *The page describes the damage and repairs to the USS Franklin during attacks in October 1944. It includes details of bomb damage, a suicide plane crash, and the ship's involvement in the Battle for Leyte Gulf.*

BODY-10 *After a bomb detonation on the flight deck, firefighting operations were initiated. Sprinklers and water curtains were turned on in the hangar, but some were damaged. Dense smoke and heat forced personnel to evacuate, but they later reentered with hose streams. The fire was confined between frames 90 and 160.*

BODY-11 *Fires broke out on the hangar and gallery decks of a ship after an explosion. The fires were eventually extinguished, but there was no gasoline vapor explosion. Some doors and bulkheads were damaged, and personnel had to evacuate due to smoke and heat.*

BODY-12 *A series of fires and explosions occurred on a ship, causing damage to various compartments. The cause of the incidents is unclear, but gasoline leaks and vapor are suspected. The ship also experienced a list to starboard, prompting counterflooding of certain areas.*

BODY-13 *The page describes the damage and flooding caused by a fire on the USS Franklin during World War II, as well as the efforts to control the fire and repair the ship.*

BODY-14 *At 0708, an enemy plane bombed the USS Franklin without radar warning, while 45 planes were aloft and 53 remained on board.*

BODY-15 *Two bombs struck the ship, causing extensive damage to the flight deck, hangar, and gallery deck structures. The first bomb detonated on impact with the armored hangar deck, creating a hole in the armored deck and blowing out light bulkheads. The second bomb detonated below the gallery deck level, causing heavy damage in the area. Aircraft fuel tanks ruptured, leading to rapid spread of fires and a gasoline vapor explosion.*

BODY-16 *During a naval disaster, multiple explosions occurred on the flight deck of a ship, causing heavy damage and fires. Efforts to control the fire were hindered by continuous explosions and limited access to sprinkler controls.*

BODY-17 *The page describes the damage caused to the hangar and flight deck of a ship during an attack. The fire suppression systems were damaged, but some sprinklers and water curtains remained partially effective. The hangar deck protected spaces below from serious damage.*

BODY-18 *Multiple rocket detonations caused significant damage to the deck and bulkheads of the ship, resulting in holes and destruction of various equipment. The hangar deck and elevator were extensively damaged, with fragments puncturing multiple decks below.*

BODY-19 *The page describes the events following explosions on the Franklin ship, including the transfer of wounded personnel, damage to machinery spaces, evacuation of personnel, loss of steering control, and assistance from the Santa Fe ship in firefighting efforts.*

BODY-20 *The page describes the efforts to rescue personnel and control fires on the USS Franklin after it was attacked during World War II. The ship was towed by another vessel and attempts were made to extinguish fires and clear smoke from the ship.*

BODY-21 *The page describes the ongoing difficulties and challenges faced by the USS Franklin after being attacked by enemy aircraft. Despite major fires and damage, efforts were made to stabilize the ship and resume operations. A significant number of casualties were reported.*

BODY-22 *The page describes the damage and repairs to the USS Franklin during World War II, including flooding, fires, and enemy attacks. The ship eventually arrived at Navy Yard, New York for further repairs.*

BODY-23 *The page discusses the types and sizes of Japanese bombs used in two different actions during the war, providing details on the damage caused and potential trajectories.*

BODY-24 *The page discusses the behavior of Franklin's ammunition during two different actions. In one action, only small caliber ammunition was present and caused sporadic explosions. In another action, heavy explosives were present and caused widespread destruction.*

BODY-25 *The page discusses conflicting reports on the sequence of explosions following enemy bomb hits on a ship. It mentions the rupture of aircraft fuel tanks and the ignition of gasoline vapor, which created a blast similar to that of high explosives. The behavior of ammunition under certain conditions is unpredictable.*

BODY-26 *The page discusses the sensitivity of explosives to detonation when hit by high velocity fragments and excessive heat. It also mentions the behavior of rockets when subjected to extreme heat and provides information about their composition and ignition process.*

BODY-27 *The page discusses the impact of unexploded Tiny Tim rockets on a ship, noting that they can be set in motion by fire and may not detonate until the fuse is armed. The rockets were observed to be projected off the flight deck and land in the sea rather than being projected out of the hangar.*

BODY-28 *The page discusses the damage caused by rocket heads and bombs during a shipboard fire. It highlights the unpredictable behavior of ammunition and the difficulty in distinguishing between damage caused by rockets and bombs.*

BODY-29 *The page describes the damage caused by fire and low order detonations in the handling rooms and hoists of twin mounts on a ship. The intense heat caused some damage, but there were no high order detonations.*

BODY-30 *The page describes damage caused by a bomb on the gun platform, including explosions in ammunition hoists and damage to switch control mechanisms. Other gun mounts and rocket motor stowage were also damaged, but there were no mass detonations.*

BODY-31 *The page discusses the condition of ammunition storage spaces during fires on ships. It states that although the ammunition was consumed by fire, the enclosures remained intact and undamaged. It also mentions that bulk stored ammunition does not mass detonate under severe conditions. The page further describes the structural damage caused by enemy attacks and emphasizes that damage above the hangar deck did not compromise the strength of the hull.*

BODY-32 *The page discusses the effectiveness of armored structures in preventing extensive damage to a ship during two actions. It also mentions the incorporation of structural sectionalization in carrier design to confine explosions and fires in the hangar space.*

BODY-33 *The page discusses the effectiveness of armored flight decks in protecting carriers from bomb attacks during World War II, using examples from British carriers. It also mentions two severe fires that the USS Franklin survived.*

BODY-34 *The fire in the hangar was controlled and extinguished using sprinklers, water curtains, and hose streams. The fire lasted until all the gasoline was consumed or washed away. The presence of power and functioning fire pumps was crucial in successfully controlling the fire. Fires below the hangar deck were difficult to combat due to dense smoke. The amount of ammunition on board contributed to the intensity and scope of the fire.*

BODY-35 *The page discusses the sectionalization of the firemain system on CV9 Class carriers during General Quarters. It suggests that dividing the system into four transverse sections is the most efficient arrangement, providing equalized distribution of pumping capacity and allowing for isolation of damaged sections.*

BODY-36 *The page discusses the issues with fire pumps and control systems during emergencies on CV9 Class carriers, and proposes solutions such as armoring the conflagration station and adding additional hose lengths at fire plugs.*

BODY-37 *The page discusses the installation of sprinkler heads in aircraft hangars to improve fire safety and control gasoline fires. It also mentions the use of foam as an effective agent for extinguishing small to moderate sized fires.*

BODY-38 *Experiments have shown that large-scale gasoline fires can be extinguished using high capacity foam and overhead sprinklers. The Bureau has authorized the installation of mechanical fog foam systems on flight and hangar decks of certain ships. Replacing sprinkler heads with fog heads is not recommended.*

BODY-39 *Fog heads have disadvantages compared to sprinkler heads, including being easily disturbed by draft conditions and prone to fouling. They are not suited for installations requiring operation at high elevations. Fixed fog systems have not been useful in controlling major fires on ships.*

BODY-40 *The effectiveness of fixed fog systems in controlling fires on ships is questioned, as investigations reveal that all-purpose nozzles may have been just as effective. The policy for installing fixed fog systems has been changed to limit their installation to watertight compartments containing gasoline piping and adjacent compartments.*

BODY-41 *This page discusses the compartments and ventilation systems in ships, as well as issues with hangar sprinkling systems. It also mentions the effectiveness of armored decks in protecting machinery spaces on carriers.*

BODY-42 *Flooding in the firerooms of a ship caused damage to boiler air intakes and uptakes, resulting in fires being extinguished and boilers being out of commission until dried out. However, there was no significant damage to pressure parts or refractory material.*

BODY-43 *The page discusses the damage caused by a fire on a ship, including water levels in the boilers and measures taken to prevent further damage. It also mentions the issue of smoke and heat entering machinery spaces during fires.*

BODY-44 *Inadequate ventilation and intense heat and smoke caused difficulties for engineering personnel on a ship during firefighting. Some improvised solutions were used, but the need for proper ventilation and protection was highlighted.*

BODY-45 *The page discusses the issue of smoke entering engineering spaces on CV9 Class carriers during battle damage. It mentions attempts to address the problem through sectionalization and providing auxiliary ventilation from the starboard side.*

BODY-46 *Insufficient supply of rescue breathers, but air-line masks proved effective in smoke-filled machinery spaces. Emergency generators failed to reconnect loads during evacuation due to short circuits.*

BODY-47 *The page discusses modifications to the casualty power system on a ship, including the removal of generators and the installation of additional risers and terminals for better power distribution.*

BODY-48 *Gasoline continued to flow from the damaged system, causing flooding and fires. The amount of gasoline present was insignificant compared to the fuel in aircraft tanks, which sustained severe fires for several hours. Salt water, not gasoline, would be forced out of the high overboard discharge if the internal float valve became stuck.*

BODY-49 *The gasoline system on carriers performed well, with the presence of gasoline in the system having minimal impact on fires. Dewatering after flooding was delayed due to concerns about gasoline fumes. Defueling planes in the hangar is important for fire safety, but current facilities are inadequate. Portable pumps are recommended for defueling due to poor performance of fixed units.*

BODY-50 *The page discusses the formation of vapor in a gasoline defueling system and proposes modifications to reduce the hazard. It also mentions the development of a new portable defueling unit and investigates the need for drain connections in aircraft fuel tanks. Additionally, it addresses the excessive time required to drain gasoline-back and purge the system, suggesting the use of proper eductors. Finally, it mentions that CV9 Class carriers tend to list to starboard when firefighting water is shipped below decks in large quantities.*

BODY-51 *The page discusses flooding issues on a ship, including the drainage of water from the hangar deck and the effects of counterflooding measures. It also mentions the displacement and stability calculations before and after damage.*

BODY-52 *The page discusses the effect of water on the stability of Franklin's firefighting systems during a fire. It estimates that around 5000 tons of water accumulated below the hangar deck, causing the ship to list at 15 degrees. The ship gradually came upright and then listed to port before returning to an even keel. The page also mentions improvements in ship design and damage control techniques.*

BODY-53 *Lessons in damage control technique were learned from the experiences of the Franklin ship, including the need for manned local control stations and additional personnel to fight fires. Ready room and flight deck personnel were endangered during accidents.*

BODY-54 *During General Quarters, pilots and aircrew should go to the second deck or below. Unauthorized personnel should not be in certain areas. A ladder was installed for escape from the Shipfitter Shop. Franklin developed a starboard list due to water and gasoline, so it is recommended to create a temporary port list in case of large fires.*

BODY-55 *The page discusses recommendations for improving firefighting procedures on a ship, including installing coamings around hatches, increasing hose supply, removing ammunition from planes, and creating aisles for movement of personnel.*

BODY-56 *The page discusses various operational issues related to firefighting efforts and safety measures on a ship, including the need for adequate ventilation, the use of flashproof covers on bunks, keeping bomb elevator doors closed, and conducting drills with rescue breathing apparatus under smoke conditions.*

BODY-57 *Repair Party personnel were unable to function due to a shortage of Rescue Breathing Apparatus. Lessons learned from Franklin's damage were effectively applied to handle similar damage on Intrepid. Holding frequent realistic drills in damage control is crucial. Franklin's failure to fully close armored hatches contributed to below deck damage.*

BODY-58 *The page discusses the material conditions and damage control measures on the CV9 Class carriers, specifically focusing on the variation in setting conditions and the need for additional flexibility. It also mentions the possibility of establishing a secondary damage control station and the importance of rescue breathers in combating fires.*

BODY-59 *The page discusses the use and limitations of rescue breathers in the Navy, including plans for an improved type breather. It also mentions the ineffectiveness of gas masks in smoke-filled compartments.*

BODY-60 *The page discusses the findings of an inspection on a ship, noting that casualties were primarily caused by fire and heat. It also mentions the use of gas masks in smoke-filled areas and their limitations.*

BODY-61 *The page discusses the dangers of gallery deck spaces on CV9 Class carriers and recommends eliminating or relocating them due to their vulnerability to explosions and fires. However, complete elimination is not practical, so some alterations have been made to relocate certain rooms to lower decks.*

BODY-62 *The page discusses damage to gallery spaces and casualties during two different actions. It also mentions the location of the Combat Information Center (CIC) and the need for improved emergency escapes on CV9 Class carriers.*

BODY-63 *The page discusses the need for improved emergency exits and access points on a ship, as well as the replacement of certain structures to prevent flooding. It also mentions the need for new materials to prevent smoke and heat from entering certain areas.*

BODY-64 *The page discusses the need for watertight hatch covers and improved drainage systems on aircraft carriers to prevent flooding and fire damage in below deck spaces.*

BODY-65 *No permanent drainage facilities will be provided for gallery deck spaces. Elevator platforms are susceptible to blast damage and future carrier designs may utilize deck-edge elevators. Sealed beam portable lanterns are the only lights partially effective in penetrating smoke-filled compartments.*

BODY-66 *The page discusses the limitations of current damage control lanterns for firefighting and the relocation of battle dressing stations on CV9 Class carriers to improve safety.*

BODY-67 *The page discusses the relocation of an emergency battle dressing station and the need for multiple aircraft jettisoning stations on carriers. It emphasizes the importance of fire control and training to combat hazards and suggests studying war damage experience from Kamikaze attacks for future significance.*

BODY-81 *A photo capturing the moment of a high order detonation on a ship's port side, showing debris in the air and firefighters running to escape.*

BODY-83 *A distant photo of a fire on the ship Franklin, with burning gasoline pouring out and men escaping through a net at the stern.*

BODY-86 *A photograph showing ammunition burning in a twin mount during a fire, with the fire on the flight deck aft almost extinguished.*

BODY-102 *Photo 35 shows damage caused by a Tiny Tim rocket on the hangar deck, with fragmentation pits visible on the STS uptake bulkhead.*

BODY-107 *Photo 40 shows the damage caused by explosions and fire in the upper handling room of a 5-inch twin mount. The hoists and center column are destroyed, and there is a hole in the outboard bulkhead.*

WASP

BODY-113 *Loss report of the U.S.S. Wasp (CV7) aircraft carrier in action in the South Pacific on September 15, 1942. The report includes details on torpedoes, damage control measures, machinery damage, fires and explosions, damage to fire main system, stowage of aircraft, improvements in gasoline systems, and design characteristics of the ship.*

BODY-114 *The page contains a list of photographs and plates depicting the damage and fire on the USS Wasp after being struck by torpedoes.*

BODY-115 *The aircraft carrier WASP was torpedoed and subsequently engulfed in flames due to gasoline fires. Despite intact engineering systems, the inability to control the fire led to the abandonment and sinking of the vessel. The fire main system failed to control the hangar deck fire, ultimately causing the loss of the ship.*

BODY-116 *The page discusses the loss of the aircraft carrier WASP due to internal fires and explosions. It highlights the need for improvements in gasoline stowage and handling arrangements for carriers. The report is based on the Commanding Officer's report and includes estimates of structural damage based on other war experiences.*

BODY-118 *The page describes the damage caused by two torpedo hits on a ship, including flooding, ruptured gasoline tanks, and shock damage to equipment. A fire also started in the hangar due to spilled gasoline.*

BODY-119 *Multiple explosions occurred on the ship, possibly caused by torpedoes and gasoline vapors. The explosions caused structural damage and fires on different decks of the ship.*

BODY-120 *Multiple explosions occurred on the ship, causing significant damage to various areas including the flight deck, bridge, hangar deck, and gun groups. The fires were intensified by loaded aircraft and ammunition. No evidence of mass detonation in bomb magazines or ready-service stowages was found.*

BODY-121 *The page discusses the damage and actions taken on the USS Wasp after torpedoes struck, including the rupture of a loop in the engine rooms, loss of pressure in the fire plugs, difficulty maintaining feed water supply, steering control changes, attempts to control fires, and eventual abandonment of the ship.*

BODY-122 *The page describes the sinking of the USS Wasp during World War II, detailing the torpedoes that hit the ship and the subsequent explosions and fires that led to its sinking.*

BODY-123 *The page discusses the size of warhead charges, measures taken to reduce list after an attack, the danger of leaving gaps in the liquid layer surrounding magazines, and the inability to control a fire on the hangar deck.*

BODY-125 The page discusses the failures and explosions that occurred during an incident involving torpedo hits on a vessel. It mentions eyewitness accounts of only two torpedoes, a possible third explosion caused by gasoline vapors, and the spreading of inflammable liquids due to tank ruptures.

BODY-126 The page describes the causes and effects of a fire in a hangar, including ignition from sparks and ruptured gasoline tanks. The fire spread rapidly among the aircraft, with detonations of projectiles and burning of wood decks. Most injuries were burns. Damage to the fire main system prevented water delivery to the fire.

BODY-127 The page discusses the condition of the fire main system on the USS WASP after it was torpedoed. It suggests that a large portion of the system remained intact, but some sections were cut off by the torpedo detonations. The ship had a well-designed fire main system with multiple pumps and loops.

BODY-128 The page discusses the damage control pumps and sprinkling systems on a ship, noting that the forward loop of the fire main was broken but could be isolated. It mentions a case where emergency means were used to extinguish fires after severe damage. There is also mention of a fracture in the fire main on another ship caused by shock from an enemy projectile.

BODY-130 The page discusses improvements in protecting gasoline stowage on naval carriers, including locating tanks within armored protection and using subdivided tanks. It also mentions the installation of protected piping mains to prevent the spread and ignition of gasoline vapor.

BODY-131 Improvements in subdivision and ventilation systems are being made to prevent the spread of explosive vapors in ships after battle damage. Gasoline stowages will be modified, voids will be fitted with eductors, inert gas will be used to reduce hazards, and electrical equipment will have explosion-proof characteristics.

BODY-132 The page discusses safety measures for the gasoline system on ships, including ventilation, fire extinguishing agents, and pumping systems. It also mentions improvements in firefighting equipment and training for personnel.

BODY-133 The page discusses the approval of schools for special training in gasoline handling, the completion of ship gasoline systems for these schools, and the establishment of firefighting schools. It also provides information on the design characteristics of WASP, including its underwater protection system. The conclusion states that WASP would have survived the damage from two torpedoes if there had been no fires.

BODY-134 The page discusses the loss of the USS WASP due to fire and highlights the need for improved firefighting facilities and precautions in handling gasoline on carriers.

NOTABLE PASSAGES

FRANKLIN

BODY-4 *1-1 This is a long report. An effort has been made to present a comprehensive summation of the many design and damage control problems which were disclosed or emphasized by the war experiences of Franklin. In addition, various pertinent war experiences of other large carriers have been considered in this report.*

BODY-5 *"The latter two cases of damage to Franklin illustrate thoroughly the ability of modern U.S. aircraft carriers to survive extensive damage from plane crashes, fire and heavy bombs. The basic design and construction of this class of carrier, which was developed prior to World War II and therefore without the benefit of war experience, is favorably reflected in the manner in which Franklin absorbed heavy damage."*

BODY-6 *"The conflagration in Franklin resulting from the action of 19 March was the most severe survived by any U.S. warship during the course of World War II."*

BODY-7 *"The ship was in Condition of Readiness III, except that all 40mm guns and all fire control stations were fully manned. This was doctrine in Franklin from sunrise to sunset, when within range of enemy shore based aircraft, as the minimum condition of readiness."*

BODY-16 *"The severity of the initial detonations and vapor explosion can be appreciated by the fact that there are only two known survivors from the hangar. All interior communications were lost except a single sound powered line from conn to steering aft from whence another sound powered line to main engine control was effective. All topside and interior general announcing systems and radio communications also failed."*

BODY-17 *"The flight deck was virtually demolished aft of the after elevator and extensively damaged forward to frame 115. From frame 115 to frame 50 wood decking was burned and deck plating was warped and buckled. There were innumerable holes ranging from small fragment holes to the largest hole just abaft the after expansion joint, frame 149, which measured roughly 60×80 feet."*

BODY-25 *"It is apparent that aircraft fuel tanks were ruptured by the initial enemy bomb blasts and fragments. Free gasoline and gasoline vapor were spread over a large area. In the comparatively confined space of a hangar, ignition of such a large volume of gasoline vapor is sufficient to create a blast which approaches in magnitude that of an explosion of high explosives."*

BODY-27 *"It is interesting to note that the rocket gained sufficient velocity in approximately 200 feet to penetrate 1¼ inch of STS although maximum velocity is not reached until after a 300 to 500 foot travel. It was reported but not confirmed, that the other Tiny Tim rocket motors performed similarly. While a number of Tiny Tims were observed to be projected off the flight deck and land in the sea, none were observed to be projected out of the hangar. It is probable that the motors were fired quickly and that the heads detonated later due to excessive heat rather than by impact. Observers reported that the appearance and eerie screaming sound of these rockets in action presented a terrifying and unnerving spectacle."*

BODY-31 *"In the action of 19 March the detonation of the two enemy bombs plus the subsequent explosions of the ship's aircraft ammunition caused very heavy damage to structure above the hangar deck, moderate damage to the hangar deck itself and only negligible damage to spaces below the hangar deck. Plates II and III detail this damage. It is important in analyzing this structural damage to properly emphasize its relation to the survival of the ship. The damage, although impressive to the average observer, and certainly long and expensive to repair, did not appreciably impair the strength of the hull girder. Since the main deck (hangar deck) is the strength deck on this class of carrier, damage, however severe, to structure above this deck, will not compromise*

BODY-32 *"The shielding effect of the armor was a principal factor in the survival of the ship. Only four large holes were blown in the armored portion and all four were caused by high order explosions in contact with or just above the deck, three by Tiny Tims and one by enemy bomb. The damage in each case to structure below the armored deck was not extensive and was limited to fragment holes in the second and third decks directly beneath the point of explosion. The fourth deck remained intact."*

BODY-33 *"The image experiences of several British carriers, which unlike our own were fitted with armored flight decks, demonstrated the effectiveness of such armor in shielding hangar spaces from GP bombs and vital spaces below the hangar deck from SAP bombs. Accordingly, the CVB Class was designed with an armored flight deck consisting of 1½-inch STS from frames 46 to 175 with a hangar deck consisting of two courses of 40-pound STS between frames 36 and 192."*

BODY-34 *"It is significant that the sprinklers and water curtains alone did not extinguish the fire but held it in check until hose streams could be applied and that even then the fire lasted until all the gasoline was consumed or washed over the side."*

BODY-35 *"It is considered that segregation of the CV9 Class firemain system into about four transverse sections is the optimum arrangement. This scheme involves the closing of a relatively small number of valves and provides a well-equalized distribution of pumping capacity. It also provides sufficient valves to permit isolation of damaged sections of the firemain system."*

BODY-36 *"In the action of 30 October, windows in the conflagration station on the hangar deck were blown in and personnel were forced to evacuate this space. Again, on 19 March, the conflagration station was damaged extensively and personnel on watch were apparently killed before they had time to operate any controls. In each instance electrical wiring to the controls was demolished. Other carriers have experienced similar casualties."*

BODY-37 *"War experience and recent experiments have shown that large-scale gasoline fires cannot be extinguished by overhead sprinklers alone, but that when the sprinkling system is augmented by sufficient hose streams such fires can be controlled and confined within definite limits. In general, once gasoline in aircraft fuel tanks is ignited and the fire develops into one of major proportions, involving say ten or more planes, it will continue to burn until all fuel is consumed despite application of water, where water alone is employed. Water, however, can be effectively used to reduce the intensity of the fire and to control its spread."*

BODY-38 *"The development and installation of these high capacity foam systems for carrier use is considered to be the greatest improvement yet achieved for the control and extinguishment of large-scale gasoline fires."*

BODY-39 *"Following the loss of several ships in the early part of the war, in which fire was the primary cause (Lexington, CV2*, and Wasp, CV7** are-outstanding examples) drastic steps were undertaken to reduce shipboard fire hazards, to improve firefighting facilities, to train damage control personnel and, in general, to make our ships more resistant to damage by fire. Rapid strides have been made in all these matters and, as a consequence, few ships have since been lost by fire."*

BODY-40 *"On Kadashan Bay (CVE76) a fire of burning gasoline from a suicide plane crash, bedding and personal gear in two living spaces on the third deck directly over gasoline stowage tanks and adjacent to gasoline pump rooms was effectively combatted by turning on the sprinkling system. The compartments were then entered with fire hoses using all-purpose nozzles."*

BODY-41 *"Despite heavy damage above the hangar deck level, Franklin did not sustain any direct structural and machinery damage in main or auxiliary machinery spaces. This has generally been true of other CV9 Class carriers which have been damaged by suicide planes and bombs.* It is attributable primarily to the effectiveness of the armored hangar deck (2½-inch STS) and to a much lesser extent to the armored fourth deck (1½-inch STS) which forms the overhead of the machinery spaces."*

BODY-43 *"After the forward firerooms were evacuated because of heat and smoke, water continued to drain into the boilers. High water marks in the furnaces indicated that water rose to a depth of 7 feet in the saturated furnaces and to about 6 feet in the superheat furnaces of boilers Nos. 1 and 3 (starboard boilers) and to a depth of about 4 feet in all furnaces of boilers Nos. 2 and 4 (port boilers)."*

BODY-44 *"In the action of 19 March, conditions were similar only much more severe. The fire was so intense and widespread that it was not possible to operate any supply blowers to engineering spaces without drawing in heat and dense smoke. Visibility was almost zero with all lights on and even sealed beam lights were of little value. Heat and smoke became progressively worse until it was unbearable and the spaces were then evacuated."*

BODY-45 *"In all these cases the port vent trunk was the source of smoke into these spaces. Later carriers, beginning with CV33, do not have the port longitudinal vent trunk but instead take ventilation supply from separate vent ducts, port and starboard."*

BODY-46 *"The air-line mask appears to be the most practical solution to the smoke problem in machinery spaces."*

BODY-48 *"The amount of this gasoline, however, at most could not exceed a few hundred gallons and its effect on the fires was insignificant when compared with the nearly 25,000 gallons contained in aircraft fuel tanks on the flight deck and in the hangar. This large amount of fuel is sufficient to sustain a severe fire for several hours."*

BODY-49 *"Doctrine on most carriers requires that all planes be defueled when their missions have been completed and they are returned to the hangar. This was the general practice on Franklin when a strike was not imminent. Other ships follow a rule of defueling all planes except fighters. War experience, particularly since the advent of suicide plane attacks, has definitely emphasized the importance of defueling all planes when stowed in the hangar."*

BODY-50 *"Investigation by the Bureau has shown that the current practice in the use of defueling pumps should be entirely revised and that minor modifications to the airplane defueling connections are required in order to reduce the hazard of vapor-locked gasoline in the suction side of the defueling lines and to increase the efficiency of the pumps."*

BODY-52 *"Franklin is an outstanding example of improvements accomplished since the early part of the war in the design and construction of ships to resist damage and in damage control equipment, technique and training of personnel. Damage Control and Firefighters' Schools and the program of education in firefighting and dissemination of knowledge of damage control through the medium of FTP-170-B and other publications have definitely proved their value. Training of the entire ship's company as well as damage control personnel in the technique of damage control, particularly in firefighting, has paid excellent dividends."*

BODY-53 *"Personnel in Ready Rooms, on Flight Deck and on Island Structure of the Franklin were endangered and in many cases injured or killed as a result of the plane crash and bomb explosion."*

BODY-54 *"In the event of a large fire on the Flight or Hangar Decks, it is recommended that the Bridge immediately develop a port list until Damage Control can take necessary measures to create a temporary list to port by movement of liquids - thus to flow water and gasoline off of Hangar Deck out of opened port side."*

BODY-55 *"Gasoline flowed down hatches to second and third decks."*

BODY-56 *"Flash proof covers were not on bunks. This resulted in a great deal of very nauseous smoke and a very persistent smoldering fire in several of the living spaces."*

BODY-57 *"The importance of holding frequent realistic drills in damage control cannot be overemphasized. Few ships actually operate hangar sprinklers and water curtains before damage, yet there is no other way to properly acquaint all hands with the function of the many control valves and the problems of draining off the large volume of water which deluges the hangar deck."*

BODY-58 *"There appears to be no uniformity in the material conditions as employed by various ships of the CV9 Class. In FTP-170-B, large aircraft carriers are listed as three-condition ships, yet from Action and War Damage Reports submitted by various ships of this class which describe setting of Conditions BAKER and ABLE, it is apparent that at least some of them operated as two-condition ships."*

BODY-59 *"In Franklin's action of 30 October, there were three men stationed in the laundry (B-429-E) when the crash occurred. Two of them donned gas masks and the third placed a wet towel over his face. Only the man with the wet towel survived. The other two men, wearing gas masks, were later found dead from asphyxiation in B-304-L."*

BODY-60 *"On the basis of the service experience it is apparent that Navy service gas masks are reasonably effective against smoke. Personnel must be thoroughly acquainted with their limitations, however."*

BODY-61 *"The gallery deck on present CV's is a death trap. It bears the full brunt of an explosion on the flight or hangar decks and catches the flames of almost any fire in these areas. Personnel in gallery deck spaces aboard Franklin at the time of attack, with few exceptions, had absolutely no avenue of escape and were doomed to certain death from the moment when the first bomb exploded."*

BODY-62 *"In the action of 19 March, gallery spaces were severely damaged and gutted by fire aft of the forward elevator from frames 50-205. Personnel casualties were heavy. Only one man escaped from CIC and air plot. Since CIC and air plot were almost directly over the center of detonation of the enemy bomb which hit forward and were subjected to severe blast and fragment attack, it is presumed that most of the personnel on watch were killed by the initial blast. Others probably were stunned initially and asphyxiated and burned before they could recover."*

BODY-63 *"Arrows painted in fluorescent paint or fluorescent buttons indicating the direction of egress as an aid to personnel groping in smoke and darkness may be installed by the ship's force as found necessary in organizing the ship for damage control."*

BODY-64 *"War experience has demonstrated that excessive accumulation of water on the hangar deck and in the elevator pits from fire-fighting systems will add greatly to the amount of water shipped to below deck spaces through open hatches, ruptured bomb elevator trunks, uptakes and air intakes and bomb and fragment holes in the hangar deck. The free surface effect of water spread over a large area of the hangar deck will also cause an appreciable reduction in the initial stability of the ship."*

BODY-65 *"As evidenced by Franklin's actions of 30 October and 19 March and by similar experiences of other carriers, the inboard airplane elevator platforms are so located as to be particularly susceptible to blast damage from explosions within the hangar spaces. The blast effect generally forces these platforms up a few feet and upon falling they assume a canted position within the flight deck opening."*

BODY-66 *"Its candlepower is but slightly less than that of a 1000-watt 115-volt 12-inch searchlight (approximately 250,000 candlepower). Higher candlepowers, it is believed, would be but little more effective in penetrating smoke dense enough to permit only a few feet of visibility with a 200,000 candlepower beam since vailing brightness of the illuminated smoke particles tends to obscure vision as beam intensities are increased."*

BODY-67 *"It is apparent from a study of the damage experiences of Franklin and other carriers that fire remains the major hazard to the safety and combat efficiency of aircraft carriers when subjected to suicide crashes and bombing attacks. Damage control efforts and training should be mainly directed toward the elimination of fire hazards and the rapid extinguishing of fire once it breaks out."*

Wasp

BODY-115 *"The loss of the WASP thus must be attributed to the fires which resulted from the ignition of gasoline."*

BODY-116 *"The circumstances involved in the loss of WASP were somewhat similar to those which caused the loss of LEXINGTON (CV2) inasmuch as both vessels had to be abandoned by reason of the raging internal uncontrolled fires followed by internal explosions. In neither case were the structural damage and loss of buoyancy and stability caused by the torpedoes sufficient to have caused the loss of the vessels."*

BODY-117 *"At 1445, While in a turn to the right, two submarine torpedoes were sighted close aboard approaching from about three points forward of the starboard beam. They appeared to have been fired from a point close to the vessel as they were still rising when sighted. They were estimated to have been between 10 and 15 feet below the surface when they struck and detonated."*

BODY-118 *"Survivor reports indicate that shock from the two torpedo hits was considerable. A ladder on the main deck about frame 20 was torn from its fastenings and hung straight down from the forecastle deck hatch. The four aircraft triced to the overhead forward fell and struck the planes parked on the hangar deck. The landing gears on all aircraft collapsed. This damage was evidently due to shock as no serious blast effects occurred in the hangar at this time."*

BODY-123 *"In this connection the magazine explosion which caused the loss of the bow of NEW ORLEANS* resulted from a torpedo detonation against the shell plating of a bomb magazine where no liquid layer was provided. Very probably the mass detonation of the bombs was caused by fragments of the torpedo and ship structure produced by the torpedo explosion."*

BODY-124 *"It appears that some chance of successfully controlling the gasoline fire on the second deck and of confining the fire in the vicinity of the gasoline tanks would have existed had the fire on the hangar deck been controlled and extinguished within a comparatively short length of time."*

BODY-125 "The first interior explosion occurred about 20 seconds after the two torpedo hits. This explosion is reported by the Commanding Officer to have been caused by a third torpedo. There were four survivors however who, as eyewitnesses, reported seeing the wakes of only two torpedoes. Vessels in the screen also reported observing only two torpedo detonations. In addition, it is believed that flooding and structural damage would have been much more extensive than appears actually to have been the case if a third torpedo had struck and detonated."

BODY-126 "In conjunction with the fires and explosions, it was reported by the Medical Officer that 99% of the injuries were burns of varying degrees of severity, mostly about the face, upper extremities and trunk."

BODY-127 "It appears probable that the starboard section of the fire main loop forward of the forward engine room was cut by the detonations of the torpedoes and that this section was not isolated, although out-out valves for the port and starboard sections of the forward loop were located in the forward engine room at frame 76."

BODY-128 "It should be noted, however, that these were not gasoline fires."

BODY-129 "The dangers of overhead stowage for planes have been recognized since the early days of the present war, and on 21 March, 1942 the Secretary of the Navy approved the design characteristics for the CVB41 class which stated, 'reserve airplane stowage to be provided, if practicable, on the same level as and adjacent to the ends of the hangar to eliminate plane stowage overhead'."

BODY-132 (n) The saltwater operating lines between the pump rooms and tanks are being reduced to a minimum in order to limit the penetration of the tank boundaries.

BODY-133 "WASP, in spite of her small size, would have survived the damage to the hull had there been no fires. Watertight subdivision, stability characteristics and reserve buoyancy were all adequate to absorb the damage from two torpedoes, located as these were, without fatal consequences."

BODY-134 "It is problematical whether the intense gasoline fires could have been controlled even with full use of available facilities."

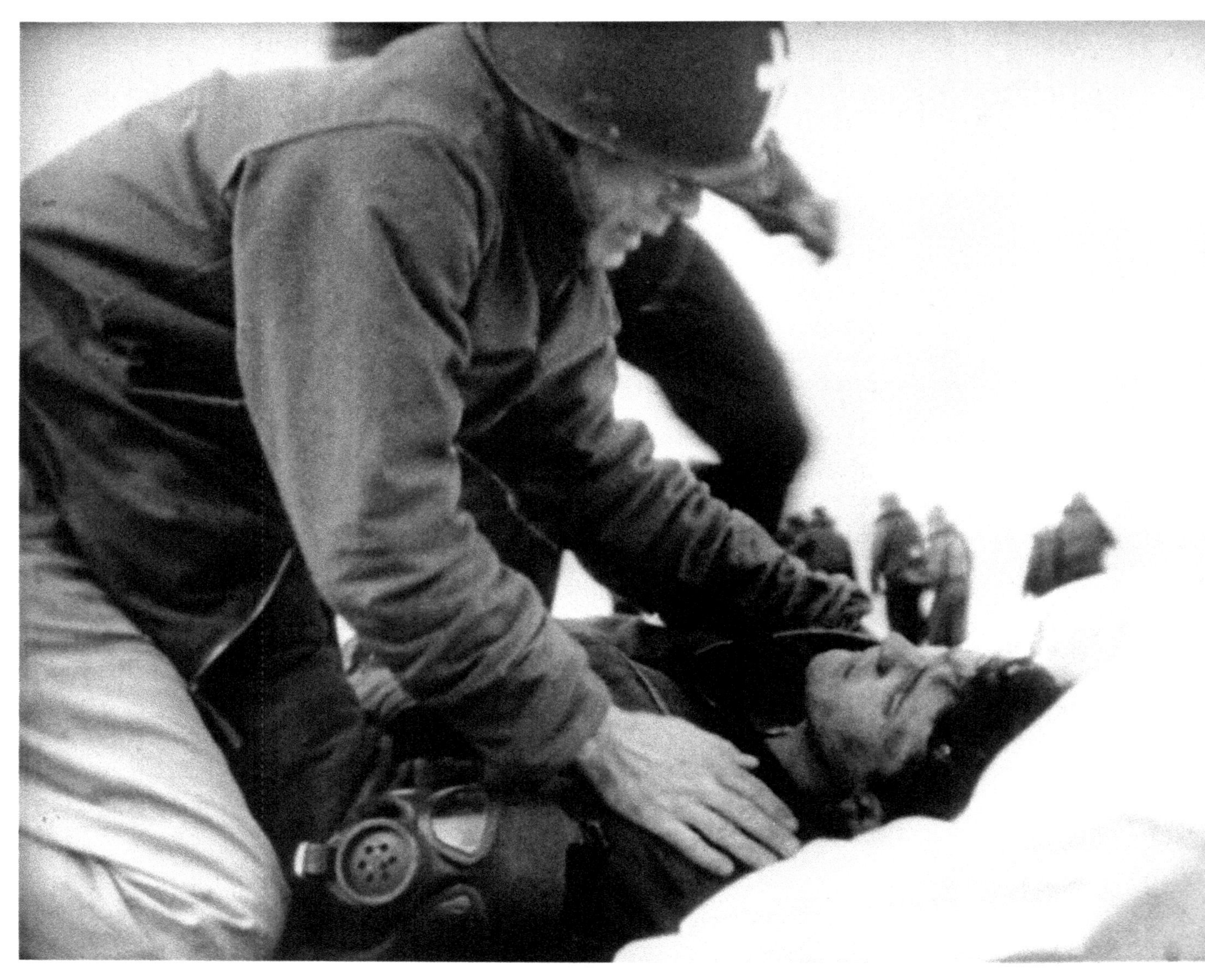

Figure 1. Lieutenant Commander Joseph T. O'Callahan, USNR(ChC) gives "Last Rites" to an injured crewman aboard USS Franklin (CV-13), after the ship was set afire by a Japanese air attack, 19 March 1945. The crewman is reportedly Robert C. Blanchard, who survived his injuries.[1]

[1] I nearly chose this photograph for the front cover. It illustrates the human cost of losing a ship at sea better than any words I could provide.—Ed.

Figure 2. The U.S. aircraft carrier USS Wasp (CV-7) burning after receiving three torpedo hits from the Japanese submarine I-19 east of the Solomons, 15 September 1942.

USS Franklin CV-13 War Damage Report No. 56

DECLASSIFIED ~~CONFIDENTIAL~~

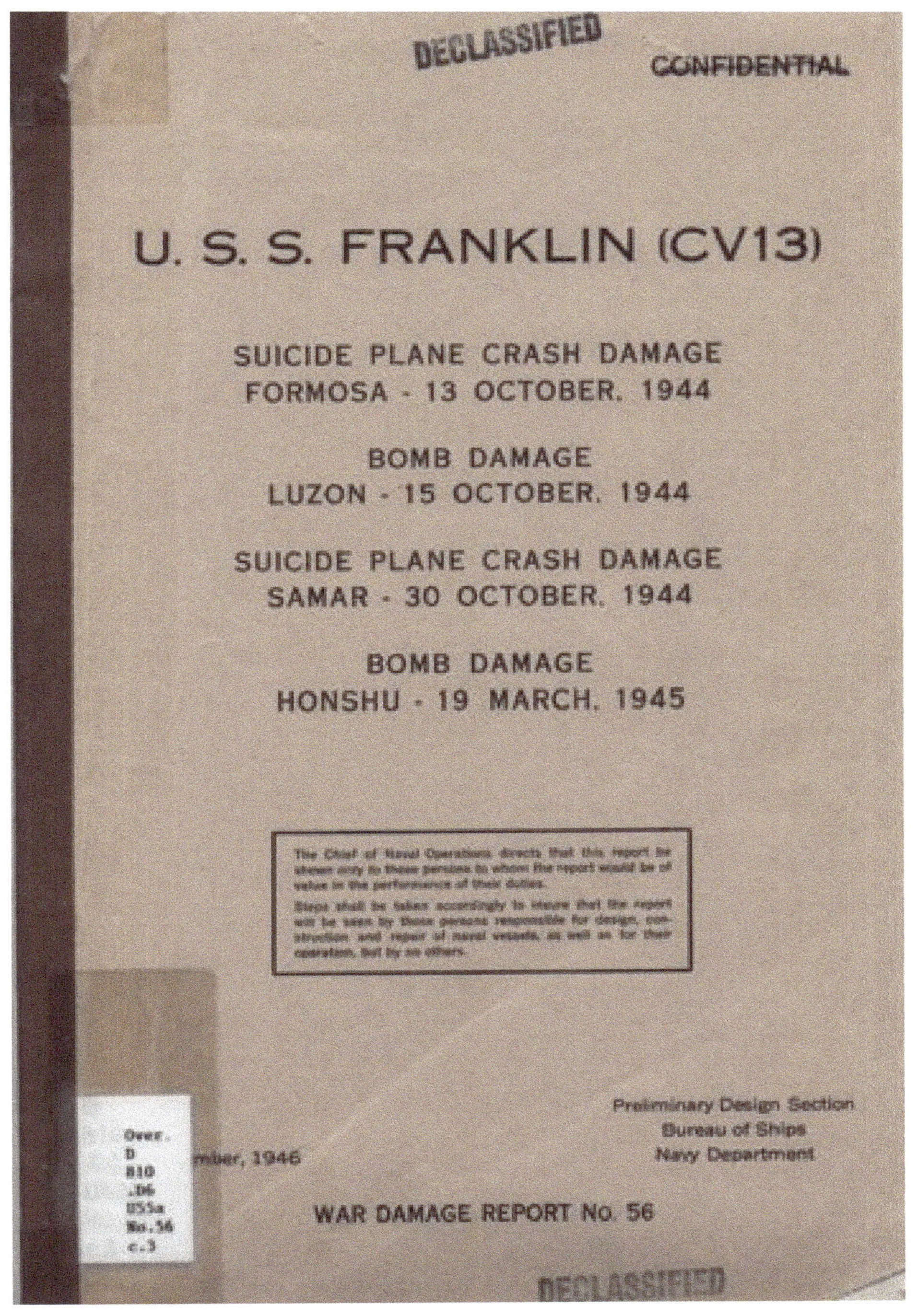

DECLASSIFIED

~~CONFIDENTIAL~~

U. S. S. FRANKLIN (CV13)

SUICIDE PLANE CRASH DAMAGE
FORMOSA - 13 OCTOBER, 1944

BOMB DAMAGE
LUZON - 15 OCTOBER, 1944

SUICIDE PLANE CRASH DAMAGE
SAMAR - 30 OCTOBER, 1944

BOMB DAMAGE
HONSHU - 19 MARCH, 1945

The Chief of Naval Operations directs that this report be shown only to those persons to whom the report would be of value in the performance of their duties.

Steps shall be taken accordingly to insure that the report will be seen by those persons responsible for design, construction and repair of naval vessels, as well as for their operation, but by no others.

mber, 1946

Preliminary Design Section
Bureau of Ships
Navy Department

WAR DAMAGE REPORT No. 56

DECLASSIFIED

Description: U.S.S. Franklin (CV13) Suicide Plane Crash Damage Formosa - 13 October 1944 Bomb Damage Luzon - 15 October 1944 Suicide Plane Crash Damage Samar - 30 October 1944 Bomb Damage Honshu - 19 March 1945

The Chief of Naval Operations directs that this report be shown only to those persons to whom the report would be of value in the performance of their duties. Steps shall be taken accordingly to insure that the report will be seen by those persons responsible for design, construction and repair of naval vessels, as well as for their operation, but by no others.

Unlimited Distribution Authorized.

15 September, 1946

Preliminary Design Section

Bureau of Ships

Navy Department

WAR DAMAGE REPORT NO. 56

Contents

V **Conclusion** 50

U.S.S. *Franklin* (CV13)

Suicide Plane Crash Damage. Formosa - 13 October 1944

Bomb Damage. Luzon - 15 October 1944

Suicide Plane Crash Damage. Samar - 30 October 1944

Bomb Damage. Honshu - 19 March 1945

Class	Aircraft Carrier (*Essex* - CV9)	Length (O.A.)	855 ft. 10 in.
Launched	14 October 1943	Beam (W.L.)	93 ft. 0 in.
Displacement (Optimum Battle Condition)	36,220 tons	Draft (Mean) (Optimum Battle Condition)	28 ft. 5 in.

References:

1. C.O. *Franklin* ltr. CV13/A16-3, Serial No. 0036, of 20 October 1944 (Report of Action on 13 October 1944)
2. C.O. *Franklin* ltr CV13/A-16-3(10-Ua), Serial No. 0038, of 23 October 1944 (Report of Action on 15 October 1944)
3. C.O. *Franklin* ltr. CV13/A12-1/A16-3(10-Ua), Seriasl No. 0039, of 31 October 1944 (Report of Action for Period 7 to 21 October 1944)
4. C.O. *Franklin* ltr. CV13/A12-1/A16-3/(10-kn), Serial No. 0041, of 4 November 1944 (Report of Action for Period 22 to 31 October 1945)
5. C.O. *Franklin* ltr. CV13/A16/L11-1(50-jd), Serial No. 0046 (No date) (War Damage Report - Action 30 October 1944)
6. P.S.N.Y. War Damage Report - *Franklin* Action, 30 October 1944
7. C.O. *Franklin* ltr. CV13/A16-3, Serial No. 00212, of 11 April 1945 (Action Report for Period 14 to 24 March 1945)
8. C.O. *Franklin* ltr. CV13/A9, Serial No. 00202 (War Damage Report - Action of 19 March 1945)
9. ComServPac ltr. FM/A9/(70-8), Serial No. 0095, of 15 April 1945 (Visitation Report on Battle Damage to Main Boilers)
10. BuMed ltr. BuMed-X-BLN-II of 17 May 1945 (Inspection of *Franklin* - Regarding Toxic Gas Problems)

11. Flag Officer Commanding, First Aircraft Carrier Squadron, British Pacific Fleet, Doc. No. 0201/10/24 of 24 May 1945
12. BuShips ltr. C-CV9 Class/L9-3(5812); NObs-81, of 30 April 1945
13. BuShips ltr. C-CV9 Class/L9-3(5812); NObs-317, of 12 May 1945
14. BuShips ltr. C-CV9 Class/S93(5812); CV9 Class/S30; NObs-1003 of 1 November 1945
15. BuShips ltr. CV9 Class/S93(5812); NObs-81, of 8 December 1945
16. ShipAlt CV891, of 18 January 1946
17. BuShips ltr. S93-3/(512); EN28/A2-11, of 22 January 1943
18. BuShips ltr. C-S93(812), of 8 February 1943
19. BuShips ltr. CVB41 Class/S93(512); CV9 Class/S93; CVE105 Class/S93; AV11-13/S03; AC18/S93, of 17 March 1945
20. BuShips ltr. CVB/S93(640-512-812); CV/S93; CVE/S93. EN28/A2-11, of 1 April 1945
21. BuShips ltr. C-CV9 Class/L-3(5812); NObs-81, of 7 June 1945
22. BuShips ltr. CV9 Class/S62(660b-5812); CV9 Class/S64; EN28/As-11, of 18 July 1945
23. BuShips ltr. CV/S15(5812); CVE/S15; CVL/S15, of 20 July 1945
24. BuShips ltr. CV/S15(812); AV/S15; EN29/A2-11, of 31 August 1944
25. BuShips ltr. CV9 Class/S15-2(512), of 12 February 1945
26. BuShips ltr. C-CV9 Class/L9-3(5812); NObs-81, of 8 June 1945
27. BuShips ltr. CV9 Class/S16-1(5812) NObs-1003, of 31 October 1945

--iii--

PHOTOGRAPHS & PLATES (AFTER PAGE 50)

--v/vii--

Section I - Foreword

1-1 This is a long report. An effort has been made to present a comprehensive summation of the many design and damage control problems which were disclosed or emphasized by the war experiences of *Franklin*. In addition, various pertinent war experiences of other large carriers have been considered in this report.

1-2 The damage sustained by *Franklin* as a result of the actions of 13 and 15 October 1944 was superficial and is included in this report only for the purpose of rendering her damage history complete. The major damage sustained in each of the actions of 30 October 1944 and 19 March 1945 demonstrates the effectiveness of bomb hits when received by aircraft carriers during the extremely vulnerable period just prior to and during periods of launching strikes. The damage sustained on 30 October is a reasonably good example of what may be expected from a suicide plane crash and subsequent fire on a carrier having a full complement of planes on board which are gassed but not armed except for small caliber ammunition. Similarly, the damage sustained on 19 March may be considered as about the maximum to be expected from fires and detonations of large numbers of bombs and rockets on the flight and hangar decks when a carrier having heavily armed, fully fueled planes aboard is hit by one or more bombs properly placed.

1-3 The latter two cases of damage to *Franklin* illustrate thoroughly the ability of modern U.S. aircraft carriers to survive extensive damage from plane crashes, fire and heavy bombs. The basic design and construction of this class of carrier, which was developed prior to World War II and therefore without the benefit of war experience, is favorably reflected in the manner in which *Franklin* absorbed heavy damage. Materiel alterations and improvements in damage control organization and technique during the war further increased the ability of this class carrier to minimize potentially severe damage. At the same time many lessons have been obtained from the experiences of *Franklin* and other cases of war damage and results of this knowledge have been and will be incorporated in existing ships where feasible and in future design and construction.

1-4 This report is based on the references, inspections of *Franklin* upon her return to this country, and informal interviews with various officers attached to *Franklin* by representatives of this Bureau.

Section II - Summary

2-1 On 13 October 1944, while engaged in aircraft strikes against the Island of Formosa, *Franklin* was damaged by a Japanese plane which made a suicide crash on the flight deck. The plane struck the ship on the port side abaft the island structure, slid across the flight deck and appeared to explode as it hit the water off the starboard beam. Only minor damage was sustained; the flight deck was gouged in several places by the propeller of the plane as it skidded across the deck and three 20mm guns were temporarily put out of commission. Necessary repairs were made by the ship's force.

2-2 On 15 October, while engaged in aircraft strikes against enemy airfields in the Manila area, *Franklin* was damaged by one bomb which struck and detonated on the after outboard corner of the deck-edge elevator. Damage to the ship was slight; the deck-edge elevator remained in operation, three planes were damaged, and a small gasoline fire in the hangar was extinguished quickly. Necessary repairs were made by the ship's force.

--1--

2-3 On 30 October, while standing by east of Samar Island to give air support to shore operations on Leyte Island, *Franklin* was damaged by a Japanese suicide plane which crashed through the flight deck. The crash and subsequent detonation of the plane's bomb load at about the gallery deck level caused extensive structural damage and started fires of great magnitude on the flight and hangar decks. The fires were fed by gasoline from plane fuel tanks. Fires and vapor explosions spread to various compartments as far down as the third deck. Adequate firemain pressure was maintained and by skillful and tenacious effort all fires were extinguished in about 2½ hours. This fire was considered the most serious conflagration any U.S. vessel had survived up to that date. After completion of repairs and authorized alterations by the Navy Yard, Puget Sound, *Franklin* was returned to service on 26 January 1945.

2-4 On 19 March 1945 while conducting air strikes against targets on the Japanese Islands of Kyushu and Honshu, *Franklin* was struck by two bombs which detonated in the hangar. This attack occurred at a most inopportune time inasmuch as a strike was being launched and 31 planes were still on the flight deck, fully gassed and armed with bombs and rockets and 22 planes were parked in the hangar, some of which were gassed and armed with rockets. Direct damage resulting from detonation of the enemy bombs was extensive in itself, but appears minor compared with the immense damage caused by subsequent fires, explosions of bombs and rockets, and water used in firefighting. Major fires raged on the flight and hangar decks and in gallery spaces for approximately ten hours. Stubborn smoldering fires in the gallery spaces and Commanding Officer's country plus recurring gasoline fires on the fantail were not completely burned out or extinguished until 22 March. During the first five hours following the initial damage, conditions were totally out of control due to the early loss of firefighting facilities and intermittent heavy explosions of 500-pound and 250-pound GP bombs loaded on the flight deck planes, some of which fell through to the hangar deck, and Tiny Tim rockets on both the flight and hangar decks. Large areas of the flight deck and hangar and gallery spaces were wrecked. All power was lost when dense smoke and heat forced engineering spaces to be evacuated. Personnel, casualties were severe. Even before the fires were extinguished and while ship's ammunition was still exploding, *Franklin* was taken in tow. Main propulsion power was regained on 20 March and the ship proceeded to Ulithi and thence to the Navy Yard, New York where complete repairs and authorized alterations were accomplished. *Franklin* was returned to service (Inactive Fleet) on 14 June 1946.

2-5 The conflagration in *Franklin* resulting from the action of 19 March was the most severe survived by any U.S. warship during the course of World War II. It is pertinent, however, to point out that the resulting damage would not in itself have caused the loss of the ship since the principal strength structure, watertight integrity and vital machinery below the hangar deck remained intact and the stability characteristics were at all times sufficient. This is principally attributable to the excellent shielding effect of the armored portion of the hangar deck. That *Franklin* was not abandoned and scuttled even though dead in the water only 50 odd miles from, the Japanese Islands of Kyushu and Shikoku, and almost untenable as a result of fire and explosions, was due largely to the courageous and determined action of the 103 officers and 603 men who remained aboard to extinguish fires and to put the ship back into operation. It was fortunate that the tactical situation permitted the ship to be taken in tow and provided with a screen and air cover until out of the immediate danger zone.

--2--

Section III - Narrative

A. Suicide Plane Crash Damage - 13 October 1944

3-1 On 13 October 1944, *Franklin* was flagship of a Fast Carrier Task Group engaged in aircraft strikes against the Island of Formosa. By late afternoon all strike missions had been recovered on board. Enemy planes had been in the vicinity of the formation during the day and the ship had gone to battle stations at 1654 on an alert from the Task Group Commander, but had been secured from this alert at 1727. The ship was in Condition of Readiness III, except that all 40mm guns and all fire control stations were fully manned. This was doctrine in *Franklin* from sunrise to sunset, when within range of enemy shore based aircraft, as the minimum condition of readiness. *Franklin* was engaged in landing eight fighters returning from combat air patrol when an attack group of five Japanese medium bombers ("Bettys") carrying torpedoes approached out of a rain squall on the port side. The altitude of the enemy planes was 50 to 75 feet and they were spaced at about one-minute intervals. No warning of the approach of these planes was received from CIC. One of the five planes was turned away by the fire of the screening destroyers and cruisers. The remaining four were taken under fire by *Franklin* at an average range of 4000 yards. Left full rudder was ordered. The first plane, on a relative bearing of 270 degrees, was hit repeatedly and set on fire but continued to approach *Franklin*. It released a torpedo at about 500 yards which passed under the fantail. Although partially out of control and burning, this plane attempted to crash *Franklin*. It struck the ship on the flight deck just abaft the island, slid across the flight deck and appeared to explode upon striking the water on the starboard beam. The ship was then brought back to base course. The second and third attacking planes were shot down by ships in the group assisted by one of *Franklin's* planes. The fourth "Betty" approached *Franklin* on the port bow and released its torpedo. It was shot down in flames just after hedge-hopping over the bow. *Franklin* immediately turned right using full rudder and backed with the starboard engine. The torpedo passed ahead within 50 feet of the bow.

3-2 Damage sustained by *Franklin* from this attack was comparatively minor and had little effect on the ship's fighting efficiency. Several gouges were made in the flight deck, frame 165, by the propeller of the plane. Two 20mm guns were damaged slightly as the plane crashed by the starboard gun sponsons on its fall from the flight deck. The barrel of a third 20mm gun was holed by shrapnel. Necessary repairs were accomplished by the ship's force.

B. Bomb Damage - 15 October 1944

3-3 Immediately following the attack on Formosa the Task Group headed south at high speed to attack enemy airfields in the Manila Bay area on Luzon. On the morning of 15 October, while east of Luzon *Franklin* was assigned the task of conducting air strikes against Nichols Field. From 0720 until the time of the enemy attack at 1046 several bogies appeared on the radar screen and all were dispersed or shot down by the Task Group's CAP. At 1046, while the first strike was out and a second deck load of planes was ready on the flight deck, two "Oscars" and one "Judy" armed with two bombs each were not tallyhoed but were first detected visually over the Task Group. Two of these planes dive-bombed *Franklin*. Only one bomb struck the ship and this detonated upon impact with the after outboard corner of the deck-edge elevator. The resulting damage was slight and the elevator

--3--

remained in operation. A small fire was started in the hangar by fragments igniting gasoline from ruptured aircraft belly tanks but was extinguished quickly by fire parties using all-purpose nozzles. One plane on the flight deck and two in the hangar were jettisoned due to damage from this hit. Necessary repairs were accomplished by the ship's force.

C. Suicide Plane Crash Damage - 30 October 1944

Plate I. Photos 1 through 10

3-4 Following the bomb damage of 15 October 1944, *Franklin* remained in the vicinity of the Philippine Archipelago as Task Group flagship. She supported the occupation of Leyte Island which commenced 20 October and participated in both the Central and Northern actions of the Battle for Leyte Gulf, 23 to 25 October. On 30 October *Franklin* was about 100 miles east of Samar Island, standing by to give air support on call to shore operations on Leyte Island.

3-5 The ship was in Material Condition BAKER; speed was 15 knots. Visibility was over 12 miles with unlimited ceiling. The wind was from the southeast, force 18 knots. The sea was light. There were about 45 fighters fueled and armed with 50 caliber ammunition spotted aft of frame 120 on the flight deck and about 45 planes, fueled but not armed, parked in the hangar. All planes were fitted with droppable wing or belly fuel tanks which were full. The ship's gasoline system was secured and purged with inert gas. There were no bombs or torpedoes in the hangar.

3-6 At 1405 launching of 12 fighter planes by catapult was commenced. At 1410 enemy planes were picked up by radar at a distance of about 37 miles. At 1412 Torpedo Defense was sounded and all anti-aircraft batteries were manned. At 1417 the destroyer which had been fueling alongside was cast off. At 1419 General Quarters was sounded. Catapulting of the 12 fighters was completed at 1420. At about 1423 Material Condition ABLE was ordered set, course was altered and speed was increased to 18 knots. Procedure in *Franklin* required separate orders for General Quarters and Condition ABLE, presumably to allow battle stations to be manned prior to closing up the ship. The attack on the Task Group was made by 6 enemy planes, either "Zekes" or "Judys," 3 of which made suicide runs on *Franklin*. The first plane missed and crashed into the sea about 20 feet from the port side abreast frame 120. At 1426 the second plane struck the flight deck at an angle of approximately 20 degrees with the horizontal somewhat to starboard of the centerline at about frame 127. This plane, with its bomb load intact, crashed through the flight and gallery decks into the hangar. The bomb, estimated to be 250 kg. GP, detonated at about the gallery deck level. At the time of the crash (about three minutes after Condition ABLE was ordered set), only 5 of the 8 repair parties had reported completion of setting Condition ABLE. Repair I (hangar), Repair V (engineers), and Repair VIII (flight deck) had not reported. The fact that Repair I was slow was to have later serious effects. The third plane approached low over *Franklin* and dropped a bomb which detonated harmlessly in the sea about 30 feet from the starboard side abreast frame 60. This plane continued on and made a suicide crash against *Belleau Wood* (CVL24).

3-7 The impact of the plane which struck *Franklin* and the detonation of its bomb load blew a hole in the flight deck approximately 12×35 feet somewhat to starboard of the centerline between frames 125 and 128. The deck was bulged upward and torn between frames 125 and 128 on the port side and between frames 121 and 130 on the starboard side (Plate I). Gallery deck spaces between frames 121 and 143 were wrecked by blast and fragments (Photos 3, 4, 5). The after airplane

elevator was forced up about 2 feet and assumed a canted position in the flight deck opening. The operating plungers were bent. This damage rendered the elevator inoperable.

3-8 Raging fires started instantly among planes on the flight and hangar decks (Photo 1). Blast and fragments ruptured aircraft fuel tanks which caused the fires to spread rapidly and to increase in intensity. Fires also were started by blast, burning gasoline, and gasoline vapor explosions which penetrated to various compartments below the hangar deck level.

3-9 On the flight deck the fire quickly spread to the other planes in the vicinity of the after elevator and engulfed 5-inch mounts Nos. 5 and 7. Hose lines which had been laid out near gassing stations prior to the damage were promptly manned by Repair VIII. Firefighters approached the fire from forward and aft. All-purpose fog nozzles were found to be the most effective equipment although some 12-foot replicators with low velocity heads were used to advantage. Solid streams were employed to wash gasoline over the side. Foam was tried but was reported to be ineffective due to high wind. The fire was confined between frames 115 and 150. Firemain pressure on the flight deck dropped slightly soon after firefighting operations started due to the fact that hangar sprinklers and water curtains had also been turned on. At this time the firemain was segregated into four sections with port and starboard cut-out valves closed at frames 68, 110 and 141. The only two fire pumps not on the line were the after Diesel pumps. These were promptly cut in and within two minutes pressure was considered adequate on the flight deck. All 14 pumps were now on the firemain. Fires on the flight deck were extinguished by 1530 except for the wood decking which continued to smolder and reignite until all fires in the hangar and gallery deck spaces were extinguished (Photo 2). Eight planes were jettisoned on the flight deck.

3-10 In the hangar, immediately after the bomb detonation, all sprinklers and water curtains in bays Nos. 3, 4 and 5 were turned on. Some of these were operated from the hangar deck conflagration station and some from the individual control booths in the hangar bays. Observation windows in the conflagration station were blown out and the personnel in this space were burned and forced to evacuate shortly after operating the sprinklers. It was later found that the No. 3 bay sprinklers and the water curtain between bays Nos. 3 and 4 were largely demolished. However, the water flowing from the ruptures was partially effective. A break in firemain riser No. 16 was isolated with difficulty because dense smoke on the second deck prohibited access to the cut-out valves. The break was finally secured by closing valves in the No. 4 fireroom and the after engineroom.

3-11 Dense smoke and heat in the hangar forced all personnel to evacuate immediately after the bomb detonation. Eighteen men of Repair I were trapped for about one hour in the shipfitter's shop on the hangar deck, B-121-E, frames 111-121, starboard. These men were evacuated by means of lines to the flight deck. It was about 30 minutes after the fire started before firefighting parties could reenter the hangar at the forward and after ends. Hose streams were then brought into service to augment the sprinklers and water curtains. Even at this time it was not possible to enter the hangar without a rescue breather. The ship's entire allowance (106) of rescue breathers was used, but many more could have been employed to advantage. Firefighting in the hangar was further handicapped by lack of vision since the dense smoke could not be penetrated by the JR-1 battle lanterns. Sealed beam lights penetrated the smoke only a few feet.

--5--

3-12 During the 30-minute interval in which access to the hangar was not possible, the sprinklers and water curtains very effectively confined the fire between frames 90 and 160. Upon return of personnel it was apparent that the sprinklers in bays Nos. 1, 2 and 5 and the water curtain in bay No. 5 were not needed and these were secured. All firefighting equipment in the hangar was employed. Foam was found to be ineffective with the sprinklers operating except in the after elevator pit where it was very effective. The 40mm and 20mm ready service lockers in the vicinity of the damaged area were sprinkled with the sprinkling system. The upper handling rooms for 5-inch guns Nos. 5 and 7 were soaked using fire hose lines. Later, other ready service stowage spaces were sprinkled for a short period to lower the temperature.

3-13 Fires in the hangar deck spaces were extinguished by 1625. Fires in the gallery deck spaces were extinguished by 1635. These latter fires were confined between frames 110-145 and were particularly difficult to combat due to restricted access to many of the spaces involved. They were finally brought under control by inserting applicators with low velocity heads and solid hose streams through holes cut in the flight deck and in bulkheads.

3-14 Apparently no gasoline vapor explosions occurred in the hangar and gallery deck spaces although there were large quantities of burning gasoline present on the hangar deck. Fortunately, there were no bombs or torpedoes in the hangar, although small arms ammunition exploded sporadically creating some physical and great mental hazard to personnel.

3-15 The initial blast demolished the No. 3 bomb elevator trunk on the starboard side of the hangar at frames 125-128. Blast and flash passed down the trunk, blew off the counterweighted door at the third deck level and started a fire in the galley, B-318-L (Photo 9). The inner (flame seal) door was open for some unknown reason, the man assigned to close it was killed by the blast at his station at the elevator controls on the third deck. The blast continued on down the trunk to the elevator pit on the fourth deck level and blew off the watertight door connecting the bomb elevator machinery space, B-435-T, with compartment B-431-E, in which are located the uptakes and air intakes for boilers Nos. 7 and 8. Trunk bulkheads were dished a maximum of 5 inches between decks.

3-16 When the crash occurred a member of Repair Party I was about to close the armored hangar deck hatch at frame 109 starboard, but was killed before he could complete his assignment. The hatch was not closed until some time later. In the meantime, water, dense smoke and burning gasoline drained down this hatch starting fires in crew's space, B-201-EL, frames 100-111 on the second deck. Water and burning gasoline then drained on down from this space through the non-watertight hatches*, port and starboard, into B-310-AL on the third deck. Smoke spread throughout numerous compartments on the second and third decks.

3-17 Dense smoke and heat caused repair party personnel to evacuate B-510-AL and B-318-L before any attempt to extinguish the fires could be made. Repair party personnel stationed in B-313-L led a fire hose through the watertight door at frame 121 to fight the fire in B-318-L, but were forced to retire because of heat and smoke. Other personnel stationed in the laundry, B-429-E, directly below B-318-L, attempted to lead a fire hose up through the hatch at frame 129 but the dense smoke and fire in B-318-L stopped them. In retiring they abandoned

---

* No covers were provided for these hatches in the original specifications for CV9 Class. See paragraph 4-94.

--6--

the hose without shutting off the water, flooding B-429-E to a depth of two feet. Fires in B-318-L, B-310-AL and B-201-EL were finally extinguished by personnel who returned wearing rescue breathers and using all-purpose nozzles. Fixed fog systems were not used in the below deck berthing compartments due to the smoldering nature of fires in bedding and lockers. These were finally extinguished by solid streams and soaking.

3-18 A brief fire started in B-324-L which was promptly extinguished by Repair III using fog from all-purpose nozzles. It is reported that this fire was started by gasoline leaking from the after elevator pit through a ruptured bellows expansion joint at the foot of the auxiliary elevator plunger casing in this space. A minor explosion, believed to have been caused by gasoline vapor from the same source, occurred in this space a few minutes later. Only superficial damage was done by both the fire and explosion.

3-19 About ten minutes after the initial damage a severe explosion centered in B-319-L. The third deck was pushed down and ruptured and the second deck dished up, both to a maximum deflection of about 12 inches. The three watertight doors of this compartment were blown out and the bounding bulkheads slightly dished. There was no evidence of fire. The explosion was apparently a gasoline vapor explosion, although the means by which vapor entered the compartment is not clear. It was possibly introduced by the supply ventilation system taking air from a blower in B-320-E; however, the ship reported that this system was secured. It is also possible that connecting doors to B-318-L or B-310-AL were open and gasoline vapor accumulated in B-319-L.

3-20 There was evidence of a third explosion, not reported by the ship, centering in the general workshop, B-313-E. The second and third decks were each dished to a maximum deflection of approximately 2½ inches at frame 115, port. Slight dishing occurred on bulkhead 111, the inboard longitudinal uptake enclosure bulkheads, frames 111 to 121 (Photo 7) and the transverse uptake enclosure bulkhead 114-1/2 (Photo 10).

3-21 During the course of damage control activity the ship took a list to starboard. When this had reached three degrees counterflooding of certain port voids was ordered. It was thought possible at this time that the list was due to underwater damage caused by the near miss on the starboard side. Due to smoke and fire conditions on the second and third decks it was impossible to check for such damage by inspection. Counter-flooding was continued until the ship reached a port list of 2 degrees at which time the damage control officer decided that the initial list had been caused by the accumulated firefighting water on the various decks. Counterflooding was then promptly secured.

3-22 The air intake ducts for boilers Nos. 7 and 8 were damaged by blast and fragments at the hangar deck level. Water from the hangar drained down these ducts and also through the damaged bomb elevator trunk into B-431-E where it overflowed the coaming around the individual air intake openings and then drained into No. 4 fireroom (Photo 8). The fireroom flooded to a little above the lower floor plates before flooding was controlled and the water extinguished boiler fires, necessitating securing boilers Nos. 7 and 8.

3-23 Dense smoke entered all engineering spaces through the supply ventilation systems. The firerooms are supplied through the port longitudinal vent trunk having intakes on the port and starboard quarters of the hangar and the enginerooms are supplied from this trunk and also from separate intakes located on the starboard side of

--7--

the hangar amidships. Supply ventilation systems were secured and exhaust ventilation systems left running which reduced the smoke somewhat. Heat was intense and although watches were changed frequently, many men were overcome. Some rescue breathers were used but were insufficient in number. Despite these difficulties, no machinery spaces were evacuated and power and firemain pressures were maintained.

3-24 As soon as the fires were under control, work was started on removal of firefighting water which had collected below decks. An appreciable percentage of this water on the third deck was removed by letting it drain through hatches and escape trunks to engineering spaces where it was pumped overboard by bilge pumps. Electric submersible pumps, handy billies, and bucket brigades were employed where water could not be directed to engineering spaces as noted above. Considerable difficulty was experienced with clogging of strainers on the submersible pumps.

3-25 Following this action, *Franklin* returned to the mainland under her own power. Complete battle damage repairs and many alterations were accomplished by the Navy Yard, Puget Sound, and the ship returned to service on 26 January 1945.

D. Bomb Damage - 19 March 1945

Plates II and III. Photos 11 through 43

3-26 On 19 March 1945 *Franklin* was flagship of a Fast Carrier Task Group engaged in conducting air strikes against targets on the Japanese home islands of Kyushu and Honshu. Launching of a pre-dawn fighter sweep to attack targets on Honshu was completed at 0557.

At 0617 the Task Group Commander ordered Condition III set on all antiaircraft batteries and Material Condition YOKE set in all ships. The radar screen was reported clear of all bogies. On *Franklin* complete Condition III was not actually set but all batteries and all fire control stations except Director II were in condition to open fire immediately. Instead of Material Condition YOKE, a modified Material Condition ZEBRA was set. This provided for 1/6 of the crew to be relieved for messing at a time and for one designated hatch (frame 109, starboard) from the hangar to the second deck to be open. The engineering plant was split with all boilers on the line. The firemain system was segregated into eight sections. The sea was calm with a 12-knot wind from about 060 degrees true. The sky was overcast with occasional breaks and low scattered clouds at an estimated height of 1500 to 2000 feet. Horizontal visibility was excellent.

3-27 At 0649 the ship was brought into the wind and speed increased to 24 knots to launch the day's first heavy strike. Launching commenced at 0657 and *Franklin's* radar screen was clear of bogies at this time. At 0705 *Hancock* (CV19) reported via TBS that her lookouts had sighted a twin engine enemy plane and at 0706 that the bogie was closing *Franklin*. All batteries, surface and bridge lookouts and sky controls were alerted. At 0708, still without radar warning, an enemy plane dived out of the base of a cloud from less than 2000 feet altitude and about 1000 yards directly ahead, made a low level "masthead" bombing run on *Franklin* and dropped two bombs. The plane was ineffectively taken under fire by one forward twin 5-inch mount and one island 40mm mount as it passed abreast these guns.

3-28 Forty-five of *Franklin's* planes were aloft and 53 remained on board, 31 on the flight deck and 22 in the hangar. Condition and spotting of planes on board was as follows:

--8--

Flight Deck

(All planes spotted aft of elevator No. 3 except 2 VB just forward)

5 VB - Gassed - Each armed with two 250-pound and two 500-pound GP bombs
14 VT - Gassed - Each armed with four 500-pound GP bombs
12 VFB - Gassed (belly tanks) - Each armed with one 11.75-inch rocket (Tiny Tim)

Hangar

11 VF - Gassed (belly tanks) - Not armed - 8 forward, 3 amidships
5 VFB - Gassed (belly tanks) - Each armed with one 11.75-inch rocket (Tiny Tim) - 3 amidships, 2 aft
4 VF - Degassed - Not armed - aft
1 VT - Degassed - Not armed - aft
1 VB - Degassed - Not armed - aft

3-29 All planes on the flight deck were turning up and one VB had just cleared the forward elevator for its take-off run. The VFB's in the hangar were spotted at the after elevator waiting to be sent to the flight deck. The forward gasoline system was secured and purged with inert gas. The after gasoline system was in operation; topping off had just been completed on the flight deck planes and three planes on the hangar deck were being topped off from the after port gasoline filling station.

3-30 The two bombs, falling in a trajectory estimated to be approximately 25 degrees from the horizontal, struck the ship almost simultaneously. The first bomb, estimated to be 250 kg. SAP containing 133 pounds of explosive, struck the flight deck to port of the centerline at frame 68, penetrated to the hangar and detonated either upon impact with the armored hangar deck or just above the deck at frame 87. There was evidence (a scarred depression in the deck) that the bomb ricocheted off the deck at frame 80. The detonation blew a hole in the armored deck (two courses of 1¼-inch STS) approximately 6×12 feet, between frames 85 to 89 just to port of the centerline. Deck plating at the periphery of the hole was deflected downward. Fragments pierced the second deck directly below the opening in the hangar deck. Light bulkheads on the second deck in way of ship's offices, frames 79-93, were blown out and distorted by blast. Blast and fragments also caused extensive damage in the hangar and to gallery deck structures. The deck of CIC and air plot, directly over the point of detonation, was riddled by fragments. The conflagration station was wrecked. The forward elevator was lifted up out of place and the plungers pulled out of their cylinders. When the platform dropped, it assumed a canted position of about 45 degrees with the flight deck and the starboard plunger punched holes in decks below to the fourth deck level (Photo 29).

3-31 The second bomb, estimated to be 250 kg. GP, struck the flight deck in the vicinity of the after elevator and penetrated to the hangar where it is believed to have detonated just below the gallery deck level and over parked planes. Because of heavy damage in this area from subsequent explosions it was not possible to identify the path and damage of this second enemy bomb. Detonation of this bomb, however, did not blow a hole in the armored hangar deck nor in the armored second deck in way of the after elevator pit. Some observers reported that the second bomb detonated a few seconds after the first one while others reported that the interval of time between the two detonations was not discernible and that the sound and shock was as if a single explosion had taken place.

--9--

3-32 Detonation of the enemy bombs in the hangar ruptured aircraft fuel tanks causing fires to spread rapidly on both the flight and hangar decks. A tremendous gasoline vapor explosion followed the initial detonations by a few seconds. Blast and flames filled the entire hangar and shot up elevator wells and out the sides of the hangar. Dense black smoke filled the hangar and enveloped large portions of the flight deck and bridge. Planes on the flight deck, which had been turning up were thrown together with their propellers cutting into one another. The severity of the initial detonations and vapor explosion can be appreciated by the fact that there are only two known survivors from the hangar. All interior communications were lost except a single sound powered line from conn to steering aft from whence another sound powered line to main engine control was effective. All topside and interior general announcing systems and radio communications also failed.

3-33 Within a short period of time, variously reported as from one to four minutes after the initial detonation, the first of a five-hour long series of heavy explosions of aircraft bombs occurred. During this period it is estimated that about 60 of the 66 500-pound bombs and about 7 or 8 of the 10 250-pound bombs which were loaded on planes on the flight deck detonated. Some planes, together with their bombs, were blown over the side without their bombs exploding. Three 500-pound and two 250 pound bombs were found unexploded in 20mm gun tubs on the port gallery walkway. Most of the bombs on the flight deck exploded on that deck, but some fell through holes and exploded in the hangar spaces. All of the 12 Tiny Tims (11.75-inch rockets) on the flight deck went off. Some were observed to leave the ship by the force of the motors, but it is believed that a majority of the rocket heads detonated on the ship. Four of the five Tiny Tims in the hangar detonated. One intact Tiny Tim rocket head was recovered in C-201-2L on the second deck. Small caliber, 20mm, 40mm and 5-inch ammunition exploded singly (low order) throughout the period of the heavy explosions and for several hours following. This ammunition was located in planes, clipping rooms, ready service boxes and upper handling rooms. No lower magazine spaces were involved. Fires, fed by gasoline and aggravated by the continuing explosions, raged unabated during the first few hours on the flight deck and in the island, gallery, forecastle, hangar and a few second deck spaces (Photos 11-18, Plate II).

3-34 At 0725 the ship was steadied on a course with the wind broad on the starboard bow. Speed of the ship was 16 knots. This served to clear smoke from the forward end of the ship and allowed firefighting personnel to enter the forward end of the hangar and to approach the fire on the flight deck. Violent explosions which were occurring at frequent intervals together with continuous low order explosions of small caliber, 20 and 40mm and 5-inch ammunition prevented firefighters from approaching close enough to have any effect in bringing the conflagration under control. Their efforts, however, were effective in preventing fires from spreading to the extreme forward end of the ship.

3-35 It is not known if the hangar sprinkler and water curtain controls were operated by the watch in the conflagration station, or whether they operated from shock or damaged circuits. The two men on watch at this station were killed and the station wrecked. Smoke and debris blocked access to a number of the third deck hangar sprinkling and water curtain control stations. Those that could be reached were turned on. Risers and lines for sprinkling and water curtain systems in the amidship section and aft in the hangar were practically demolished. In the hangar as far forward as frame 44, risers and overhead lines were broken and cut by fragments in many places. Sagging of overhead structures from excessive heat caused considerable overhead piping to be carried away.

--10--

3-36 Although fire pumps in machinery spaces were kept on the line, it was not possible to maintain adequate pressure on all sections simultaneously. At the time of the hits, the firemain system was divided into eight sections. Since it could not be determined which sprinkling and water curtain systems in the hangar were still effective, segregation of the firemain loop system was not changed. Immediately after the initial damage the two Diesel fire pumps aft and one of the two Diesel fire pumps forward were started (the other one forward could not be started). With all available fire pumps operating, a large volume of water was discharged into the hangar through damaged and undamaged risers, sprinkling and water curtain lines. At least some of the sprinkling and water curtains were partially effective despite the damaged piping. Water curtain No. 2 and sprinkling bay No. 1 were reported in operation and these aided in preventing spread of the fire forward. The firemain loop below the fourth deck remained intact. Salt water flushing lines were ruptured on the second deck in several compartments and this contributed to flooding lower spaces.

3-37 One of the two known survivors from the hangar reported that immediately following the initial bomb detonations he and another man led out a hose line from the aviation repair shop and started forward. Before they had gone more than a few feet, the first subsequent explosion (gasoline vapor) knocked both men to the after end of the hangar. Only one of these two men escaped and he made his way to the fantail, later transferring to a ship alongside. The only other survivor from the hangar was manning a gasoline system telephone on the starboard side at frame 164. He was also blown to the after end of the hangar and subsequently made his way to the fantail.

3-38 The flight deck was virtually demolished aft of the after elevator and extensively damaged forward to frame 115 (Photo 25). From frame 115 to frame 50 wood decking was burned and deck plating was warped and buckled. There were innumerable holes ranging from small fragment holes to the largest hole just abaft the after expansion joint, frame 149, which measured roughly 60×80 feet (Photos 21, 22, 23, 26).

3-39 Although the armored hangar deck (two courses of 1¼-inch STS from frame 26 to frame 166) was ruptured in four places and extensively scarred and warped, it was very effective in protecting spaces below from serious damage. As noted in paragraph 3-30, the enemy bomb which detonated at frame 87 blew a hole roughly 6×12 feet in the armored deck and fragments pierced the second deck directly below. The other three holes in the armored deck were identified as having been caused by the detonation of Tiny Tim rockets. In each of these three cases the characteristics of the ruptures indicated that the rocket heads had detonated while lying flat on the deck. One hole, about 3×6 feet, was blown in the deck at its junction with the inboard 2-inch STS bulkhead of the uptake space at frame 93. The armored bulkhead of the uptake space was torn open from the deck to a height of about 4 feet. The uptake and air intake for boilers Nos. 1 and 3 were damaged but not beyond limited use. A second hole about 6 feet in diameter was blown in the deck at frame 100 at its junction with the bulkhead of forward uptake space (Photo 35). This hole extended into the after part of the uptake space and demolished the starboard air intakes for boilers Nos. 2 and 4. The uptakes for boilers Nos. 2 and 4 also were severely damaged. A third Tiny Tim rocket is believed to have detonated on the deck at frame 146, aft and outboard of the after starboard corner of the after elevator. A hole about 6×6 feet was blown in the deck. Light bulkheads of the crew's berthing space, B-127-L, frames 131-145, were demolished. In all three cases small fragment holes were blown in second deck plating directly below the Tiny Tim detonations. Fragments from the Tiny Tim detonation at frame 146 also pierced the third deck at frames 144-150, starboard. The fourth deck directly below was depressed, frames 146-148, but was not penetrated.

--11--

3-40 A large hole, roughly 14×18 feet, was blown in the center of the hangar deck, frames 176-180. In this area, deck plating is only 3/8-inch medium steel. The force of the detonation of the bomb (or bombs) which caused this hole also blew a hole 8×12 feet in the second deck and fragments punctured the third deck directly below.

3-41 In addition to large holes described above, there were many smaller fragment holes in the hangar deck, particularly aft of frame 166 where the deck plating is not STS. The deck was depressed over a major portion of the area aft of frame 155 and in some places between frames 80 and 155.

3-42 The after elevator was wrecked in a somewhat similar manner to the forward elevator except more extensively (Photo 28). Both plungers were pulled out of their cylinders and the elevator platform was riddled by fragments and assumed a canted position. All bomb elevators, barriers, arresting gear and similar fittings on the flight deck aft of frame 50 were destroyed. The mast support for the SC-3 radar antenna fractured and in falling smashed both the SC-3 and SM antennae. The foremast which supports antennae for the YE homing beacon, SG and BK radars and VHF radio gear was fractured at the radar platform level. This mast tilted inboard but was prevented from hanging down by its starboard wire stay (Photo 14).

3-43 At about 0900 *Miller* (DD535) came alongside *Franklin* and took off the flag officers and their staffs. At about 0930 *Santa Fe* (CL60) took station about one hundred feet off the starboard bow of *Franklin* and started transferring seriously wounded personnel by trolley. *Franklin* slowed to about eight knots to facilitate the transfer.

3-44 Immediately following the initial detonations dense smoke entered all machinery spaces through supply ventilation ducts. As subsequent explosions ruptured decks down to the fourth deck, fire and smoke entered numerous other spaces below the hangar deck level through the holes and open accesses. Ventilation supply systems to machinery spaces were secured and exhaust blowers were left operating in an effort to reduce smoke. Lack of supply ventilation made these spaces extremely hot and increased the difficulties of personnel remaining on watch. The detonation of the Tiny Tim rockets adjacent to the forward uptake spaces on the starboard side of the hangar demolished the air intake for boilers Nos. 2 and 4 and damaged the air intakes for boilers Nos. 1 and 3. As stated in paragraph 3-39, the uptakes for these boilers were also damaged. Water from hangar deck firefighting drained down these ruptured air intakes and extinguished fires in all four forward boilers. Blast passing down the air intakes and uptakes for boilers Nos. 2 and 4 caused a flareback. No. 2 fireroom was evacuated at about 0900.

3-45 At about 0930, smoke and heat conditions in the remaining firerooms, enginerooms and auxiliary machinery spaces had become progressively worse. Men were collapsing at their stations and permission was granted to evacuate all machinery spaces. Throttles were set for 8 knots, all boiler fires extinguished, turbo-generators secured and all firerooms, enginerooms and auxiliary machinery, spaces were evacuated. The two emergency 250 KW Diesel generators started automatically, but circuit breakers on the main switchboards opened, presumably from overload conditions, and both generators operated without load until machinery spaces were remanned approximately six hours later. All firemain pressure was lost except one section forward supplied by Diesel fire pump No. 11. The two after Diesel pumps in the refrigeration machinery room, C-614-E, frames 166-176, were started and left running when this space was evacuated because of smoke. This compartment subsequently flooded and it is not known how long the pumps operated.

--12--

3-46 Steering control was lost within a few minutes after evacuation of machinery spaces, and at about 1015, the ship lost all headway and started to swing. This made it impossible for *Santa Fe* to maintain her position and she cast off all lines and backed away. At 1050, after *Franklin* had regained a steady heading, *Santa Fe* came in on the starboard bow slamming into actual contact where she was held by use off her engines. The remainder of the seriously wounded and excess personnel were then transferred and *Santa Fe* cleared the side about 1225. While alongside, *Santa Fe's* firefighting personnel directed hose streams with some effect on the gasoline fire amidships, on the fires in 5-inch 38 caliber twin mount No. 7 and 40mm ready service boxes and bulwark stowages amidships. In the meantime, *Hickox* (DD673) and *Miller* (DD535) approached the stern, picking up *Franklin's* personnel from the water en route and took off wounded and other personnel trapped on the fantail.

3-47 By about 1000 a list to starboard, caused by firefighting water accumulating on the hangar deck and decks below, had increased to eight degrees and continued increasing approximately one degree every ten minutes. Heavy explosions were continuing but at longer intervals.

3-48 At 1115 *Pittsburgh* (CA72) was ordered to take *Franklin* in tow even though severe fires and major explosions still continued. *Pittsburgh* lay to on the port bow and passed over an 8-inch manila messenger followed by the towing wire. No power was available at the forecastle winches or anchor windlasses on *Franklin* and the process of heaving in the messenger by manpower was long and tedious. Power handling was finally arranged by using a winch on *Pittsburgh*. Upon receipt of the towing wire aboard *Franklin*, it was attached to the starboard anchor chain, the anchor shackle having been burned off with a portable burning outfit, and 90 fathoms of chain was eased out of the chain locker by slow ahead movement of *Pittsburgh*. Arrangements were completed and towing commenced at about 1400. *Franklin's* rudder was shifted from amidships to three degrees right by the hand positioning gear. Course was difficult to maintain at first because *Franklin* tended to sheer to port and sail to windward, dragging the *Pittsburgh*'s stern around. Towing speed averaged 4 to 5 knots by 2400.

3-49 All major explosions ceased about 1300 although 40mm and smaller caliber ammunition continued to explode intermittently. Fires on the forward end of the hangar deck had been extinguished and those on the flight deck, after part of the hangar deck, in the forecastle, gallery island and second deck spaces had either burned out or had become confined and were gradually being brought under control. Holes were cut in the flight deck and hose streams directed on fires in gallery deck spaces below (Photo 25). The only fire pressure on the ship at this time, as noted in paragraph 3-45, was from the one Diesel fire pump forward. Accompanying destroyers assisted in fighting fires on the fantail and after part of the ship.

3-50 At about 1300, the engineer officer together with several officers and men, all wearing rescue breathers, made their way to the forward auxiliary machinery space and found the forward emergency 250 KW Diesel generator still running at no load. No. 1 distribution board was stripped and the generator cut in on the board. This provided power for lights and ventilation blowers for machinery spaces which were started immediately. Ventilation blowers for the third deck spaces were also cut in. Thus, clearing the dense smoke from the third deck accesses and from the machinery spaces was begun. An inspection of the machinery spaces made with use of rescue breathers disclosed that the forward fireroom was not usable. By 2100 it was possible to remain in No. 3 fireroom without rescue breathers and preparations were made to light off.

--13--

3-51 *Franklin's* difficulties were not limited to the immediate problems on the ship itself, for two more enemy air attacks were attempted during the afternoon of the 19th. At 1254 a "Judy" made a fast low level attack on *Franklin* from starboard, but was taken under fire by the starboard screen and its bomb exploded short, about 200 yards on the starboard quarter. At 1435 an enemy aircraft closed in but was driven off by anti-aircraft fire and splashed by the CAP.

3-52 By night all major fires were extinguished. However, there were many smoldering fires scattered throughout the damaged area and these rekindled and flared up at frequent intervals. During the night one bad fire on the fantail was extinguished with the aid of one of the screening destroyers.

3-53 During the early afternoon the starboard list of the ship had stabilized at about 13 degrees. The ship was about three feet down by the stern. Counterflooding of various port voids was started with the intention of bringing the ship back to about 5 degrees starboard list and holding it there. Some of the flooding control stations for the port voids were accessible only with rescue breathers. The ship responded gradually to the list control measures, but due to either an overestimate of the amount of counterflooding necessary or lack of coordination in the counterflooding effort, together with the appreciable reduction in GM due to the free surface of the accumulated firefighting water on the various decks, the ship came upright about 0000 and slowly listed to port, finally coming to rest with a 9 degree port list at about 0400 on 20 March.

3-54 At about 2230 No. 5 boiler was lighted off and as soon as sufficient steam pressure was available No. 3 turbo-generator was started and cut in on the No. 3 main distribution board. By 0100 on 20 March warming up of main engines was commenced. At 0715 boiler No. 7 was lighted off and at 0815 boiler No. 8 was lighted off. The after main engines were then turned over and brought up to 56 turns. With this assistance *Pittsburgh* was able to increase speed of the tow to 6 knots. By 1000 steering control had been regained and four boilers were on the line. A report was made to the Task Unit Commander in Guam that the ship was ready to make 15 knots and permission was requested to cast off the tow, which was granted. Speed was gradually increased to 14 knots and at 1233 the tow line was cast off and *Pittsburgh* hauled clear.

3-55 Beginning with the dawn of 20 March, personnel remaining on the ship started extricating and burying bodies, clearing wreckage and dewatering flooded spaces. About 300 men and 100 officers were fit for duty out of a total of 603 men and 103 officers on board.

During the day *Franklin* worked up to 18 knots steaming on the four after boilers and with the two after main engines operating normally. Because of flooding and damage to air intakes and uptakes forward boilers were not lighted off. Auxiliary steam from the after plant was cross-connected to the forward engines which were turned over to reduce the drag of the outboard propellers. One gyro-compass was in operation. All but a few stubborn fires in gallery deck spaces, Captain's cabin and some of the lower deck spaces aft were out. These fires were not completely extinguished until the morning of 22 March.

3-56 Pumping of the after port damage control voids which had been flooded to offset the initial starboard list was commenced on 20 March. On the morning of 22 March the list had been reduced to 6 degrees to port and, although it had been planned to retain about 5 degrees port list in order to pocket damage water, as off-center and free surface water was removed the ship gradually came to an even keel. Water was removed by submersible pumps and bucket brigades and also by draining it to lower compartments where it was pumped out through the main drainage system. This was a slow and tedious task.

--14--

3-57 Several more attempts were made by the enemy to attack *Franklin*. At 1452 on 20 March an undetected "Judy" made a low level bombing run from the direction of the sun on the starboard bow of *Franklin*. This plane was not taken under fire by the screen on its approach. At about 1500 yards *Franklin's* forward island 40mm quad mount, in manual operation and local control by a volunteer crew, took the plane under fire. This caused the plane to pull up and swerve at the dropping point, with the result that its bomb crossed the flight deck, barely missing the port deck edge, and exploded in the sea about 200 feet off the port quarter. At 1151 on 21 March a "Betty" made a low level bombing run on the starboard quarter of *Franklin*, but it was splashed by the CAP about eight miles distant. At 1205 on the same day another bogey was splashed by CAP before it could close for an attack.

3-58 While en route to Ulithi speed was worked up to approximately 20 knots. The work of extricating and burying bodies and clearing of wreckage on the hangar deck was progressed as far as possible. The antenna of the forward SG radar was salvaged from the wrecked fore-topmast, mounted on the platform on the port side of the smokestack, hooked up by jury-rigged cables to the flag plot SG radar receiver and then connected to the remote plan position indicator on the navigating bridge. This improvised unit is reported to have operated reliably and accurately. Additional guns were restored to operation and some of the more essential power, light and telephone cables were repaired or replaced with casualty cables. *Franklin* arrived at Ulithi on 24 March and after a brief period proceeded to the mainland, arriving at Navy Yard, New York, on 26 April 1945.

IV DISCUSSION

A. Enemy Bombs

4-1 In the action of 30 October it is believed from the amount of resulting damage that the plane ("Zeke" or a "Judy") which crashed through the flight deck carried one 250 kg. GP bomb containing from 211 to 330 pounds of picric acid or Type 98 explosive. Bombs of this size and type were used by the Japanese in the last year of the war.

4-2 In the action of 19 March, the Japanese bomb which struck forward at frame 68 is believed to have been a 250 kg. Type 99, No. 25 SAP bomb containing approximately 133 pounds of tri-nitro-anisol. The diameter of the body, 11.5 inches, corresponded to the holes in the flight and gallery decks which were slightly larger. A scarred depression in the hangar deck at frame 80, directly in line with the path of the bomb, indicated that it glanced off the deck and detonated one or two feet above the deck at about frame 87 (Plate II). Because of the low angle of trajectory of this bomb, estimated by observers to be 25 degrees from the horizontal, it is considered probable that it ricocheted upon impact with the deck. The nature of the hole blown in the armored hangar deck and the fragment scars in the plating around the hole indicate that detonation occurred fairly close to the deck. It is significant that fragmentation damage from this bomb was extensive above the armored deck; whereas, below the armored deck the damage was minor. Several medium-sized fragment holes occurred in the second deck, but these are believed to have been caused by fragments from the armored deck.

4-3 No conclusive evidence is available on which to estimate the path followed and the size and type of the second Japanese bomb in the action of 19 March. Reference (g) reported that it struck the flight deck at about frame 133 and detonated above the hangar deck at about frame 142. This would place the point of entry at the forward end of the after elevator and the center of detonation in way of the after

--15--

end of this elevator. The after elevator, which was at the flight deck level at the time of the hit, was lifted bodily upward pulling the hydraulic plungers out of their cylinders. The damage and final position of the elevator platform, guides and plungers indicate that a heavy explosion occurred somewhat to port and below the platform at about the gallery deck level. The fact that this bomb detonated well above the armored hangar deck without hitting any heavy structure indicates that it probably was a general purpose bomb similar to the type used in the action of 30 October. The Japanese developed and used several GP bombs weighing approximately 250 kg. with explosive charges varying from 211 to 330 pounds, any one of which might have been employed in this attack. Plate II shows an estimated path of this second bomb, based on actual flight deck damage. While it is somewhat unusual for a Japanese plane to carry two bombs of two different types in the same load, this may have been done in the attack on 19 March.

B. Behavior of *Franklin's* Ammunition

4-4 In the action of 30 October fortunately all planes were armed only with small caliber ammunition. There were no bombs, torpedoes or rockets in the hangar or on the flight deck. Fires caused the small caliber ammunition to explode sporadically with low order detonations. It is noteworthy that hose streams were used promptly to flood the upper handling rooms for 5-inch guns Nos. 5 and 7 when water pressure to the sprinkling system failed due to fracture of a firemain riser in the hangar.

4-5 In the action of 19 March, a large amount of heavy explosives was present on the flight and hangar decks. The stunning effect on personnel and destruction of material by the initial and subsequent explosions, together with the rapid spread of fire, prevented any appreciable jettisoning of planes and ammunition. The location of the ammunition involved in fires and explosions is summarized as follows:

Loaded on Aircraft	
Flight deck	66 - 500-pound GP bombs 10 - 250-pound GP bombs 12 - 11.75-inch rockets (Tiny Tims) Machine gun ammunition on 31 aircraft
Hangar	5 - 11.75-inch rockets (Tiny Tims) Machine gun ammunition on 5 aircraft
Stowed in Ship Spaces	
Flight deck and island	5-inch/38 cal. ready service powder cans and projectiles inside twin mount No. 7 20 and 40mm ammunition in ready service rooms and bulwark stowages*
	5-inch/38 cal. powder cans and projectiles in twin mount No. 7 upper handling room*
Gallery deck	5-inch/38 cal. powder cans and projectiles in ready service stowages for single mounts Nos. 6 and 8 on port quarter*

---

* Believed to be full or nearly so.

--16--

Loaded on Aircraft	
	5-inch rocket motors in B-0206-1/2, frames 131-133 port*

Hangar deck	20 and 40mm ammunition in ready service rooms and bulwark stowages* 20 and 40mm ammunition in ready service stowages*

4-6 Information as to the sequence of explosions immediately following detonation of the enemy bombs is conflicting. It was reported by some observers that ship's ammunition began to detonate immediately following the enemy bomb hits. Others reported that there was a lapse of time between the initial detonations and the first subsequent heavy detonation of from 1 to 4 minutes. Also, there was divergency of opinion as to whether the first detonation of ship's ammunition occurred in the hangar or on the flight deck. One phenomenon which was supported by reliable observations was the heavy vapor explosion which immediately followed the enemy bomb hits. Almost simultaneously, flames accompanied by dense black smoke filled the hangar and shot up the elevator openings and from the sides of the hangar, quickly enveloping large portions of the flight deck and bridge. It is apparent that aircraft fuel tanks were ruptured by the initial enemy bomb blasts and fragments. Free gasoline and gasoline vapor were spread over a large area. In the comparatively confined space of a hangar, ignition of such a large volume of gasoline vapor is sufficient to create a blast which approaches in magnitude that of an explosion of high explosives. Dense black smoke is characteristic of burning gasoline vapor. This vapor explosion may have confused some observers who presumed it to be a bomb explosion, accounting for the varying estimates as to when the ship's ammunition started to explode.

4-7 Reference (g) noted that after the detonation of the enemy bombs and the subsequent vapor explosion and envelopment of the flight deck and island by dense smoke, the ship was slowed and turned to starboard, then turned to port and standard speed ordered. The enemy bomb which struck forward did not start a fire on the flight deck. The bomb which struck aft ruptured some aircraft fuel tanks on the flight deck and consequently started fires simultaneously on the flight and hangar decks.

4-8 The subject of ammunition behavior under predetermined conditions is still not predictable to any degree of certainty. It cannot be safely concluded that a set of conditions which produce detonation in one case will continue to so behave, and vice versa. However, it is known that projectiles and bombs loaded with high explosives are generally subject to detonation, both high** and low** order, under the following two conditions:

1. When subjected to the roasting effect of high temperatures applied for an appreciable interval of time. Instances are on record in which the filler has burned out completely without detonating; in others, detonations have occurred after heating periods as short as three or four minutes and in some cases only after appreciable lengths of time.

---

* Believed to be full or nearly so.

** A "high order" detonation is an exothermic reaction in an explosive charge, occurring under certain conditions of temperature and pressure, which causes it to break down instantaneously into gaseous components without evidence of any unconsumed remainder and which has the maximum damaging effect upon its surroundings for the given weight of explosive. Where a part of the charge noticeably fails to contribute to the reaction, and the effect on surroundings is of relatively lesser degree by a considerable margin, the detonation is described as "low order" detonation.

--17--

1. When hit by high velocity fragments. Sensitivity to detonation in this case is dependent upon the velocity of the fragment, the temperature of the fragment, the wall thickness of the projectile or bomb and the characteristics of the explosive itself.

4-9 The reported lapse of time between the initial bomb hits and the first ship's aircraft bomb detonation of from 1 to 4 minutes indicates that the cause of this first subsequent detonation was due to excessive heat from the gasoline fires applied to one or more of the bombs loaded on the flight deck aircraft. Sporadic explosions of bombs continued for about five hours. Some bombs detonated high order while others exploded low order. It is believed that excessive heating caused the majority of the separate explosions although some were possibly caused by fragment impact either by the smaller caliber projectiles or from other explosions. Some explosions were more severe than others. One particularly heavy explosion was noted by several observers at 0952. Some bombs in detonating no doubt caused almost instantaneous detonation of adjacent bombs by fragmentation impact, thus accounting for some explosions being heavier than others.

4-10 As stated in paragraph 4-5, the five Tiny Tim (11.75-inch) rockets were initially the only bombs in the hangar. Several GP bombs, which had been loaded on flight deck planes, later dropped into the hangar through bomb holes in the flight deck. Little information is available as to the behavior of assembled rockets when subjected to extreme heat. The 11.75-inch rocket head is a standard 500-pound SAP bomb containing 150 pounds of TNT. The rocket motor contains 148 pounds of smokeless powder consisting of 4 grains of solventless extruded ballistite of cruciform section, 60 inches long. The assembled round is 123 inches long and weighs 1288 pounds. Its velocity at 70 degrees F. is approximately 800 feet per second relative to the launching aircraft. The head, or bomb proper, has the normal characteristics of a pressure-arming base-fused projectile. The propellant charge in the motor will ignite spontaneously at approximately 325 degrees F. It is therefore a reasonable conjecture that the propellant charges ignited within a matter of seconds after being engulfed by fire. Arming of the base fuse, which occurs in two stages, cannot be completed until 0.1 second after the end of acceleration. Firing of the inertia type base fuse requires a fairly heavy impact. One unexploded rocket head was recovered in C-201-2L. The motor casing lodged in the after 1¼-inch STS bulkhead of the after elevator pit (Photo 37). This rocket is presumed to have been loaded on one of the fighter bombers which was parked forward of the after elevator on the hangar deck. The fighter bomber was headed aft, hence the rocket was pointed aft. The rocket traveled a maximum of 200 feet before striking the STS bulkhead. The head continued a distance of about 50 feet after passing through the bulkhead leaving the motor casing lodged in the bulkhead at an angle of about 45 degrees with about one-half of the body on each side. The forward section passed through the bulkhead and exploded, flaring out the forward 2 feet of the casing. The grid was found in the elevator pit, having been blown out of the after end of the motor. Apparently, the intact rocket was still accelerating when it penetrated the armored bulkhead, and hence, arming of the fuse had not yet been completed; otherwise it is believed that impact would have initiated fuse action and detonated the head. It is interesting to note that the rocket gained sufficient velocity in approximately 200 feet to penetrate 1¼ inch of STS although maximum velocity is not reached until after a 300 to 500 foot travel. It was reported but not confirmed, that the other Tiny Tim rocket motors performed similarly. While a number of Tiny Tims were observed to be projected off the flight deck and land in the sea, none were observed to be projected out of the hangar. It is probable that the motors were fired quickly and that the heads detonated later due to excessive heat rather than by impact. Observers reported that the appearance and eerie screaming sound of these rockets in action presented a terrifying and unnerving spectacle.

--18--

4-11 The general question of extra hazard from rockets naturally presents itself and it is pertinent to determine how much worse, if at all, the Tiny Tims behaved than the aircraft bombs. The sensitivity of the Tiny Tim head when subjected to heat or fragment impact is the same as that of any other similar SAP bomb. The magnitude of detonation will, therefore, be similar to that of SAP bombs containing the same quantity of explosive. However, since the propellant action of the Tiny Tim motor can be initiated by a temperature as low as 325 degrees F., it certainly can be stated that an engulfing fire will almost invariably set a Tiny Tim in motion. In an open area, such as a flight deck, the Tiny Tim would most likely be catapulted overboard prior to detonation of its head, with but little damage to the ship as compared with that which would occur if the rocket remained in the fire area and detonated by roasting or fragment impact. Even in an enclosed space such as a hangar, motion of a Tiny Tim followed by impact with ship's structure should not necessarily result in detonation of the head until arming of the fuse has been completed. This was strikingly demonstrated by the action of one of the Tiny Tims on *Franklin*, as noted in paragraph 4-10. It is also entirely possible that motion of the rocket will take it out of the fire zone and thus prevent detonation of the head by excessive heat. It does not appear that Tiny Tims are any more hazardous than bombs when located in shipboard fire areas, as far as detonation of the explosive heads is concerned. It does appear, however, that the general fire hazard of Tiny Tims is greater than bombs due to the relative ease with which propellant action can be initiated.

4-12 As described in paragraph 3-39, three holes in the armored section of the hangar deck are believed to have been caused by detonation of rocket heads. In each case it appeared that the head detonated high order while lying flat on the deck. Holes varying in size from 3×6 feet to 5×6 feet were blown in the 2½-inch STS deck. These three holes and the hole made by one of the enemy bombs were the only holes blown in the armored portion of the hangar deck. The large hole in the unarmored deck plating at the after end of the hangar described in paragraph 3-40 was caused by one or more GP bombs falling through holes in the flight deck.

4-13 It was not possible to distinguish damage caused by rockets from that caused by the GP bombs in the after end of the hangar and on the flight deck. Damage to ship's structure could not be identified from the known spotting of planes before the damage as the planes were blown about by the initial detonations and some, with their bomb loads intact, were thrown off the ship.

4-14 The unpredictable behavior of the ammunition involved is demonstrated by the fact that bombs exploded sporadically over a five-hour period even though all of the bombs were engulfed in the conflagration. This is partially attributed to the varying degrees of temperatures to which the bombs were subjected. While some GP bombs undoubtedly detonated high order, it appears that at least 40 per cent exploded low order for about this percentage was found with nose fuses intact and fillers consumed. In most cases, base fuses were missing (Photo 36).

4-15 It was reported in reference (g) that "all topside 5-inch, 40 and 20mm, ready rocket, aircraft ammunition lockers and ammunition in all gun mounts aft of the bridge exploded." A large hole in the flight and gallery decks on the port quarter was reported by *Franklin* to have been caused by the explosion of ready service 5-inch ammunition. Another area of extensive damage to the flight deck port side, aft of the deck-edge elevator, was described by some observers as attributable to the explosion of 5-inch rocket ammunition. It was also reported that the 5-inch ready service locker on the starboard side of the flight deck at frame 130 exploded, throwing heavy and burning debris over the island structure and also over the after part of *Santa Fe* alongside. This presumably refers to the upper handling room for 5-inch/38 cal. twin mount No. 7 which was involved in fire. Despite these reports there was no evidence of other than scattered low order detonations although intense powder fires occurred in many areas.

4-16 The door in the inboard bulkhead of the upper handling room for No. 7 twin mount, gallery deck, frames 128-131, port, was left loosely dogged, by personnel when they evacuated this space and it was reported that this permitted fire to enter. In view of the intense fire in this vicinity, however, the contents might have ignited anyway. Ammunition in this handling room and in the gun mount above did not ignite until late in the fire. Both spaces eventually were gutted by fire (Photo 19) and damaged by low order detonations of ammunition, but there were no high order single or mass detonations. A three-cornered tear was blown in the outboard 3/4-inch STS-bulkhead of this handling room. The after bulkhead (3/4-inch STS) split at a vertical weld and was torn away from the deck at the bottom. The deck (3/4-inch STS) was dished downward and torn in way of the after port ammunition hoist. The center column and hoist in this handling room were demolished by fire and low order explosions of ammunition in the hoist and in racks. Interior fixtures showed signs of very intense heat. One intact 5-inch projectile was found lodged between a stiffener and the outboard bulkhead with its nose fuse, missing and filler consumed. (Photos 39-43). The upper handling room for No. 5 twin mount, located on the flight deck, frames 121-125, port, was not damaged by fire.

4-17 The two powder and projectile hoists servicing 5-inch/38 cal. twin mount No. 7 leading from the lower handling room on the starboard side, frame 135, second deck, to the upper handling room, frame 130, gallery deck, were damaged by low order detonations of powder and projectiles within the hoists caused by the intense heat of the surrounding fires. It should be emphasized that in neither case did fire or flash travel back down the hoists to the ammunition in the lower handling room. The major damage to the hoists occurred between the main and forecastle deck levels. The projectile hoist sustained the most extensive damage. The 3/4-inch STS trunk of this hoist was blown open at the curve just below the forecastle deck level. The upper end casing was blown off at the bolted connection to the upper adapter. Both the sprockets and the chains were demolished at the top. At the lower end of the hoist, the upper right edge of the watertight cover was bent out as if the door at this point had not been dogged down. Lower-end electric power and control installations and hydraulic piping, all below the main deck, were not damaged except by water. The powder hoist received less damage than the projectile hoist. The 3/4-inch STS trunk of that hoist remained intact but four portable covers, one above and three below the forecastle deck, were blown off. The watertight cover on the upper end casing was blown off. Sprockets and chains were damaged beyond repair. All deck and bulkhead connections remained intact (Photos 39-43).

4-18 The large opening in the flight deck and those in the forecastle and gallery decks in way of the gun platform for the 5-inch/38 cal. single mounts Nos. 6 and 8 on the port quarter, apparently were caused by a 500-pound GP bomb which dropped down from the flight deck and detonated at about the gallery deck level (Photo 30). There is no upper handling room for these guns. Ammunition is brought from the lower magazines to these single mount guns via ammunition hoists and is stowed in ready service boxes on the gun platform which is open to the weather. These boxes were still intact although gutted by fire. Ammunition in the upper ends of hoists for single mounts Nos. 6 and 8 burned and exploded low order. Hoist No. 6 begins on the port side, first platform, frame 165, and terminates on the gun platform, gallery deck, port side, frame 186. Heat from the fire ignited the ammunition within this hoist at the top and several projectiles exploded low order blowing off portable plates between the gun platform and the forecastle deck. The upper switch control mechanism and switches were damaged beyond repair. The

--20--

chain was twisted and jammed in the upper part of the hoist. The chain track of the upper section, the STS casing, and sprockets and ball bearings were damaged slightly. It should be noted that only ammunition at the top exploded. Hoist No. 8 begins on the first platform, frame 165, port side, and terminates on the gun platform, gallery deck, frame 193, port side. Damage to the upper section was similar to that sustained by hoist No. 6. One STS side plate split at the welded joint to the STS casing just below the gun platform and two upper portable plates were blown off. The roller chain was twisted and jammed. Sprockets and ball bearings were damaged slightly. The watertight cover of the upper casing was blown off. The upper switch control mechanism and switches were damaged by fire and explosions. Here again, only ammunition in the upper section was involved. No fires or damaging effect were transmitted to the lower magazine spaces in either case.

4-19 5-inch single mounts Nos. 2 and 4 and twin mounts Nos. 1 and 3 were damaged slightly by fire and water. Fires, however, did not occur inside the gun mounts or upper handling rooms.

4-20 5-inch rocket motors were stowed in two improvised spaces on the port side of the gallery deck. One space, frames 131-133 (formerly crew's shelter B-020-1/2) had about 200 rounds of rocket motors which were consumed by fire but there was no indication of mass detonation. The light bulkheads bounding this space were intact and undamaged except by fire. The other 5-inch rocket motor stowage, frames 12-14 (formerly A-020-1/2 M) was not involved in a fire and all motors were thrown over the side early in the action.

4-21 40 and 20mm ammunition in ready service rooms, clipping rooms and ready service boxes on the hangar, gallery and flight decks burned and exploded low order singly but there was no evidence of high order or mass detonations of this ammunition. The ammunition in stowage spaces involved in fires was consumed and in some instances melted down to a mass of brass and steel, yet the enclosures were intact and relatively undamaged other than by fire. For example, 40mm ready service room C-204-M, frames 178-181 on the port side of the gallery deck was initially about full of ammunition. During the action, all ammunition was consumed by fire and the cases and projectiles were split open, yet the explosions were of such low order that no damage occurred to ship's structure other than by fire. Other 40 and 20mm stowage spaces on the hangar deck were found in a similar condition.

4-22 War experience has shown that 20 and 40mm ammunition stored in bulk will not mass detonate under the most severe conditions of blast and heat. Where fires have occurred in this type of ammunition the most adverse reaction has been intermittent low order detonations as on *Franklin*. In the majority of cases the sprinkling systems for the ready service spaces involved were not operable either because of blast and fragment damage to piping or by loss of firemain pressure in the zone of damage. In some cases control valves could not be reached. Accordingly, in the interest of simplifying construction and reducing unnecessary weight, splinter protection and sprinkling systems for 20mm and 40mm ready service rooms will be omitted in future construction and the removal of the same from ships in service will be accomplished as weight compensation for military alterations of high importance.

--21--

4-23 In the action of 30 October the structural damage caused by the suicide crash of the enemy plane and the detonation of its bomb at about the gallery deck level was extensive but not critical. A hole approximately 12×35 feet was blown in the flight deck somewhat to starboard of the centerline between frames 125 and 128. Gallery deck spaces between frames 121 and 143 were wrecked by blast and fragmentation. No damage was done to the armored hangar deck (Photos 2, 3, 4, 5).

4-24 In the action of 19 March the detonation of the two enemy bombs plus the subsequent explosions of the ship's aircraft ammunition caused very heavy damage to structure above the hangar deck, moderate damage to the hangar deck itself and only negligible damage to spaces below the hangar deck. Plates II and III detail this damage. It is important in analyzing this structural damage to properly emphasize its relation to the survival of the ship. The damage, although impressive to the average observer, and certainly long and expensive to repair, did not appreciably impair the strength of the hull girder. Since the main deck (hangar deck) is the strength deck on this class of carrier, damage, however severe, to structure above this deck, will not compromise the strength of the hull.

4-25 That damage below the armored hangar deck was comparatively minor is directly attributable to the effectiveness of its armored portion, two courses of 1¼-inch STS, plug-welded together, between frames 26 and 166. The shielding effect of the armor was a principal factor in the survival of the ship. Only four large holes were blown in the armored portion and all four were caused by high order explosions in contact with or just above the deck, three by Tiny Tims and one by enemy bomb. The damage in each case to structure below the armored deck was not extensive and was limited to fragment holes in the second and third decks directly beneath the point of explosion. The fourth deck remained intact. The hangar deck was scarred and depressed in innumerable places, indicating that many heavy explosions occurred which, in the absence of armor, would undoubtedly have caused much greater damage.

4-26 In both actions, armored structures in the hangar and island were subjected to severe blast and fragment attack. On 30 October, STS plating inboard of the uptakes in the hangar effectively defeated blast and fragments, while unarmored air intakes and the after bomb elevator trunk were demolished by the blast. On 19 March, although armored bulkheads in the hangar inboard of the uptakes and air intake space, frames 93-100, starboard, were torn open, presumably by Tiny Tim rockets, it is apparent that the STS protection served an important function in preventing more extensive damage. There was evidence of fragment attack in way of the uptake space, frames 110-121, starboard, yet no fragments penetrated the inboard STS bulkhead. The 3/4-inch STS plating on the deck of CIC and air plot in the gallery, frames 82-89, which was installed during the Navy Yard availability following the damage of 30 October, defeated about 90 per cent of the fragments. These two compartments were almost directly over the center of detonation of the forward enemy bomb. Armored bulkheads and doors in the island structure effectively defeated the majority of fragments and to a large degree made the island stations tenable throughout the action. STS protection of spaces above the hangar deck level has been carried out even more extensively in the CVB Class carriers.

4-27 As a result of study of damage sustained by various British carriers prior to our entry into the war, two important departures from traditional U.S. Navy carrier design were incorporated in the CVB Class, then still under development. HMS *Illustrious* in an action

--22--

off Malta on 1 January 1941 was hit by several bombs, three of which detonated in the hangar spaces. Large fires swept fore and aft among parked planes thereby demonstrating the desirability of attempting to confine the limits of such explosions and fires by structural sectionalization of the hangar space. On the CVB Class the hangar was therefore divided into five compartments separated by 40 and 50-pound STS division bulkheads extending from the hangar deck to the flight deck, each fitted with a large door suitable for handling aircraft. It is hoped that this sectionalization, in conjunction with sprinkler and fog foam systems, will effectively prevent fires from spreading throughout the hangar spaces, as occurred on *Franklin* on 30 October and 19 March. The image experiences of several British carriers, which unlike our own were fitted with armored flight decks, demonstrated the effectiveness of such armor in shielding hangar spaces from GP bombs and vital spaces below the hangar deck from SAP bombs. Accordingly, the CVB Class was designed with an armored flight deck consisting of 1½-inch STS from frames 46 to 175 with a hangar deck consisting of two courses of 40-pound STS between frames 36 and 192. Although none of the CVB Class carriers were completed in time to take part in war operations, the effectiveness of armored flight decks against Kamikaze attacks was demonstrated by various carriers attached to the British Pacific Fleet. Reference (k) reports two such interesting cases. The *Victorious* was struck by three Kamikaze aircraft, two of which ricocheted off the armored flight deck and over the side, causing no important damage. The third carried a bomb which detonated at frame 30 starboard at the butt of the 3-inch flight deck armor with 1½-inch "D" quality (equivalent to HTS) steel. It does not appear that the Kamikaze actually struck the ship. The bomb detonation, however, depressed the 3-inch deck slightly but did not tear it open. On the other hand, the 1½-inch "D" quality deck plating was ripped open over a total area of about 25 square feet. Two days were required for temporary repairs, at the conclusion of which the ship was fully operational. HMS *Formidable* was hit by two bombs, the first of which struck and detonated on the flight deck 9 feet to port of the centerline at frame 79, directly over a deep bent and at a juncture of three armored plates. The armored deck was depressed over an area 24 feet long and 20 feet wide. Maximum depression was 15 inches. Adjacent bents spaced 12 feet forward and aft of the point of impact were slightly impressed. A hole 2 square feet in area was blown in the 3-inch deck. Three fragments penetrated downward through the ship into the center boiler room. The damage in this boiler room, which was not described, temporarily reduced speed to 18 knots. The second bomb struck and detonated on the centerline of the flight deck at frame 94. The 3-inch deck and deep bent directly below the point of impact were depressed about 4½ inches and one rivet was knocked out. However, the ship was fully operational within about 5 hours, including flight operations.

D. Fires and Firefighting

4-28 *Franklin* has experienced two of the worst fires which any U.S. warship has survived, the fire of 30 October 1944 being exceeded in severity only by that of 19 March 1945. Each attack was a surprise and occurred when conditions on board ship favored the enemy for inflicting severe damage.

4-29 The action of 30 October was an excellent demonstration that a fire of tremendous proportions and intensity in the hangar and on the flight deck can be controlled and extinguished when firefighting facilities remain operable and are utilized properly. In this action the firemain system was segregated into four sections with cut-out valves closed at frames 68, 110 and 141. All pumps were on the line except the two after Diesel pumps and these were cut in within a few minutes. Personnel were forced to evacuate the hangar almost immediately following the plane crash because of heat and dense smoke. Hangar sprinklers and water curtains, which were turned on promptly, effectively confined the fire in the hangar between frames 90 and 160 until damage control personnel could re-enter the hangar 30 minutes later. Although overhead sprinkling piping of bay No. 3 was damaged extensively and one water curtain in way of the major damage was demolished, water flowing from fractured piping was partially effective in controlling the fire. Thirty minutes after the initial damage, firefighters wearing rescue breathers re-entered the hangar and brought hose streams to bear on the fire. The hose streams combined with sprinklers and water curtains extinguished the fire in approximately 2 hours. It is significant that the sprinklers and water curtains alone did not extinguish the fire but held it in check until hose streams could be applied and that even then the fire lasted until all the gasoline was consumed or washed over the side. The fact that power was maintained and all fire pumps remained in operation throughout the period of fire-fighting was an essential factor in the successful control and final extinguishment of this fire. In spaces below the hangar deck fires were not particularly severe, but were difficult to effectively combat because of dense smoke. Non-automatic fixed fog systems were not employed. Instead, spaces were entered and fires extinguished using hose streams with all-purpose nozzles.

4-30 In the action of 19 March there were considerably fewer planes on board with a correspondingly smaller total amount of gasoline in fuel tanks to feed the fire. The after gasoline system which was in operation at the time of the hits contributed somewhat to the intensity and scope of the fire by the amount of gasoline in lines above the hangar deck level, some or all of which drained out into the hangar via damaged filling lines. It is to be noted that no part of the gasoline system below the hangar deck level was involved in the fire. The significant difference between circumstances on 19 March and those of 30 October was in the amount of heavy ammunition loaded on the planes. In the earlier action only small caliber ammunition was involved; in the later action the flight and hangar decks were veritable arsenals. It is believed that, except for the ammunition involved, the fire of 19 March could have been brought under control and extinguished promptly. As it was, however, hangar sprinklers and water curtains were demolished before they had an opportunity to produce an appreciable effect on the fire. Explosions of ship's aircraft bombs restricted damage control measures to the area at the extreme forward end of the flight deck and the early loss of power further reduced firefighting water capacity to the output of one Diesel pump forward with a limited number of hose lines. It is a credit to *Franklin's* personnel that, with so little equipment remaining in operation, they persistently combatted such a conflagration in the face of continuing heavy explosions of large bombs. Credit is also due to ships which came alongside and assisted with hose streams.

4-31 It is a matter of interest that at the time of the damage of 30 October the procedure of sectionalizing the firemain system at General Quarters provided for four (4) sections, yet in the action of 19 March the procedure at General Quarters provided for eight (8) sections. This latter system required closure of approximately 31 valves. During the Navy Yard availability following the damage on 30 October additional stop valves and risers were installed, which appreciably improved the system by relieving some overloaded risers, providing better distribution of pressure and additional means of segregating damaged sections. There appears to be no uniformity among various ships of this class as to sectionalizing the firemain system at General Quarters. One vessel divided the system into ten self-sufficient sections, each fed by

--24--

one motor or turbine-driven pump, with the four Diesel-driven pumps kept as standby units. Another ship employed a system of six sections. With the additional valves and risers authorized for CV9 Class carriers, it is possible to divide the firemain system into fourteen sections, although this is neither desirable nor practical. There is a mean between minimum and maximum sectionalization which will result in the most efficient utilization. Maximum segregation imposes unequal pump capacity and unequal pressures on certain sections. It is uneconomical in that all pumps must be operated continuously during General Quarters, and in the event of a major fire, the capacity of one pump is frequently not sufficient to meet the demands of the section it serves. Based on a study of carrier war damage experience, it is considered that segregation of the CV9 Class firemain system into about four transverse sections is the optimum arrangement. This scheme involves the closing of a relatively small number of valves and provides a well-equalized distribution of pumping capacity. It also provides sufficient valves to permit isolation of damaged sections of the firemain system. Continuous operation of all pumps would not be necessary and added pumping capacity could be obtained from any other section by opening a small number of valves. Rupture of secondary risers or lines in any one section would not cause as marked a reduction in pressure as in systems where more than four sections are employed. The firemain system was divided into four sections on both *Bunker Hill* (CV17) and *Intrepid* (CV11) when these vessels were damaged by extensive fires on 11 May 1945 and 25 November 1944, respectively. Despite the fracture of several sprinkler risers in the hangar, ample pressure was maintained in both cases.

4-32 When *Franklin's* main machinery spaces were evacuated in the action of 19 March at about 0930, power to all electric and steam-driven pumps was lost. The two after Diesel-driven fire pumps were left running when the pump and refrigerating machinery room was evacuated, although it is doubtful whether they continued to run very long due to subsequent flooding of this space. Even had these pumps continued to operate, their output would have been of no use since, due to smoke, explosions and confusion, sectionalization of the firemain loop was not changed to divert pumping capacity to the intact forward system. The service performance of the one forward Diesel fire pump which operated continuously for 18 hours, the last 14 of which were unattended, is worthy of mention. It was unfortunate that the other forward Diesel pump could not be started, for its output would have been of great value.

4-33 In the action of 30 October, windows in the conflagration station on the hangar deck were blown in and personnel were forced to evacuate this space. Again, on 19 March, the conflagration station was damaged extensively and personnel on watch were apparently killed before they had time to operate any controls. In each instance electrical wiring to the controls was demolished. Other carriers have experienced similar casualties. To prevent this casualty, the Bureau authorized by reference (1) the armoring of the conflagration station with 3/4-inch STS plating, installation of vision blocks in lieu of present fixed windows and enclosing of electric sprinkling control cables in a 3/4-inch STS trunk between the hangar deck and the armored portion of the conflagration station.

4-34 War experience on this and other ships of the class has demonstrated that 100 feet of hose at fire plugs in the hangar, gallery walkways and on the flight deck was not sufficient to reach fires. Reference (m), therefore, authorized an additional 50-foot length of hose at each of these fire plugs. This additional length is to be stowed on a separate rack in the vicinity of the fire plug and connected when required.

--25--

4-35 In several instances of hangar fires on CV9 Class carriers, damage control personnel have been trapped in the shipfitters' shop, frames 111-121, starboard. This occurred on *Intrepid*, *Bunker Hill*, and twice on *Franklin*. The shipfitters' shop has remained tenable even during severe conflagrations in the hangar due to the protection afforded by armored bulkheads of the uptake space directly inboard. To provide means to attack hangar fires from this shop, reference (l) authorized installation of a fire plug therein with two 100-foot lengths of hose. In addition, this reference authorized the installation of two 2½-inch fire plugs with Wye-Gate fittings on top of the pilothouse level; one plug to be located at approximately frames 103-104 and the other approximately frames 91-92, with two 50-foot lengths of fire hose to be provided at each plug. In order that the portable P-500 gpm gasoline-driven fire pump may be used to supply water to these two fire plugs in the event of loss of pressure or damage to risers or firemains below the hangar deck, the above letter also authorized the installation of a capped hose fitting in the branch risers to the island to which the discharge hose from the P-500 pump can be connected.

4-36 The substantial increase in allowance of droppable aircraft fuel tanks for carriers of the CV9 Class introduced a critical problem in that stowage for these tanks was provided in the overhead of the hangar in such a compact arrangement that the efficiency of the hangar sprinklers was greatly reduced. To correct this condition, reference (1) authorized the relocation of the sprinkler heads so as to project below the tank stowages. The heads were installed in a pendant position which decreased the efficiency of the water pattern and it was further found, by tests conducted at the Firefighters School, Navy Yard, N.Y. on 29 June 1945, that the space between the sprinkler heads and the overhead was not protected. In the tests, burning gasoline vapors accumulated in this space above the water spray and caused flash backs which re-ignited vapors in areas where the fire had been extinguished. The installation of sprinkler heads both above and below the tank stowages was, therefore, authorized by reference (n).

4-37 War experience and recent experiments have shown that large-scale gasoline fires cannot be extinguished by overhead sprinklers alone, but that when the sprinkling system is augmented by sufficient hose streams such fires can be controlled and confined within definite limits. In general, once gasoline in aircraft fuel tanks is ignited and the fire develops into one of major proportions, involving say ten or more planes, it will continue to burn until all fuel is consumed despite application of water, where water alone is employed. Water, however, can be effectively used to reduce the intensity of the fire and to control its spread. The time required to bring major hangar fires under control when the full facilities of sprinklers, water curtains and hose streams have been utilized promptly and effectively has varied from approximately one to four hours. For example, in *Franklin's* action of 30 October and *Intrepid*'s action of 25 November 1944, fires in the hangar burned for about 2½ hours. On *Bunker Hill*, fire in the center of the conflagration burned for approximately four hours. Fires of this duration inevitably result in extensive damage to wiring, equipment and structures in the hangar and above.

4-38 Foam has been effectively used by carriers in extinguishing isolated small fires involving gasoline in one to three or four planes, spillage from operational mishaps and accumulation in elevator pits. A few ships have reported that it has been used effectively on moderate sized gasoline fires. For example, *Enterprise* (CV6) reported:

"This ship has found liquid (mechanical) foam, as utilized in duplex pressure proportioners and directly in pickup nozzles, to be the fastest and most efficient agent for control of gasoline fires."

Although foam continues to be the most effective material known for extinguishing gasoline fires, it has not been successfully employed on large-scale hangar and flight deck fires principally because of the lack of facilities to produce and apply foam in the large quantities necessary.

4-39 Recent experiments in the use of high capacity foam (foam in large quantities) have indicated that large-scale gasoline fires can be extinguished rapidly and positively using a newly developed pressure ejection method. This method produces and applies foam in large quantities through specially designed fog foam nozzles. Additional tests have shown that when foam is applied on a large-scale gasoline fire in combination with overhead sprinklers, the foam blanket is reduced in thickness but not destroyed and remains reasonably effective. Thus, the most practical method of combatting hangar and flight deck fires is utilizing a combination of sprinklers and foam in large quantities. Accordingly, the Bureau authorized by references (o) and (p) the procurement and installation of individual riser type high capacity mechanical fog foam systems (also referred to as "deluge" systems) for the flight and hangar decks of all CV's, CVB's, CVL's and CVE's. These systems are now being installed as ship availabilities and supply of materials permit. The design for the CV9 Class provides for nine foam stations on the flight deck (actually eight on each gallery walkways and one on the flight deck in way of the island), each with a 2½-inch hose connection, and ten foam stations in the hangar, each with one 3½-inch and one 2½-inch hose connection. Each station is equipped with sufficient hose, fog foam nozzles and stream shapers. Six of the ten hangar deck stations are equipped with swivel type monitors. The cutout valves for the hangar deck stations may be operated locally or by remote control from the second deck using reach rods. Foam liquid is injected into risers feeding these stations by ten proportioner units widely dispersed on the second deck, each equipped with a 300-gallon tank of foam liquid and a pump. An additional 300-gallon supply of foam is provided in cans at each pumping station for refill as necessary. The overall weight of the new equipment totals approximately 33 tons. **The development and installation of these high capacity foam systems for carrier use is considered to be the greatest improvement yet achieved for the control and extinguishment of large-scale gasoline fires.**

4-40 It has been suggested that the present commercial type sprinkler heads in the hangar deck overhead sprinkling system be replaced with fog heads. It was stated that equal or greater cooling with substantially less water may be accomplished with fog heads as compared with commercial type sprinkler heads. Replacement of the present sprinkler heads is not recommended for the following reasons:

1. Commercial sprinkler head discharge characteristics, so far as the type of spray produced and the discharge pattern are concerned, are not greatly affected by reduced firemain pressures. When firemains and sprinkling system supply piping are

damaged to the extent that pressures drop below 60 psi the fog production qualities of fog heads are severely depreciated.

2. Type L-11 fog heads discharge 60 gpm at 100 psi. Spacing requirements being approximately equal, fog heads discharge amounts of water practically equal to sprinkler heads. Advantages claimed for less water accumulation and less secondary damage with fog heads are not altogether true.

--27--

1. Fog discharged from fog heads is finer and lighter than the spray produced by sprinkler heads. The lightweight fog discharge pattern is subject to considerable disturbance by draft conditions prevalent in hangar spaces, produced by open roller doors, blast and forced ventilation. Closer spacing of fog heads to produce heavier fog patterns results in the formation of larger water droplets, water consumption is increased and additional piping and fittings are required which results in increased weight.
2. Fixed fog installations are subject to fouling of the small holes in fog heads by marine growth, scale and any substances which may be transferred in the supply piping to such systems.
3. Standard commercial sprinkler heads deliver a coarse spray with considerable cooling capacity from 18 feet elevation. Fog heads, L-11, or any other type producing fine sprays or fogs, are not suited to installations requiring operation at such an elevation for the reasons given above. On the other hand, the coarse spray of the sprinkler head is not as much affected by air currents and a consistent spray pattern is delivered to cool the fire area and beat down the flames. Access to the fire area can then be gained by fire parties with foam and fog hose lines to complete extinguishment.

4-41 Following the loss of several ships in the early part of the war, in which fire was the primary cause (*Lexington*, CV2*, and *Wasp*, CV7** are-outstanding examples) drastic steps were undertaken to reduce shipboard fire hazards, to improve firefighting facilities, to train damage control personnel and, in general, to make our ships more resistant to damage by fire. Rapid strides have been made in all these matters and, as a consequence, few ships have since been lost by fire. The extensive installation of non-automatic fixed fog systems resulted from efforts to improve firefighting facilities on carriers. The primary purpose was to enable firefighters to progressively approach the center of a conflagration by connecting hose lines to individual systems and advancing compartment by compartment. Inspections of many battle damaged carriers and interviews with ships' officers have not revealed a single case where fixed fog-systems have been useful in controlling major conflagrations.

Actually, only a few ships have reported using these systems. On *Belleau Wood* (CVL24), a berthing compartment was sprinkled for a few minutes when it was discovered full of smoke; however, it was later found that the smoke was caused by a temporary short circuit and that actually there was little evidence of fire. On *Chincoteague* (AVP24)*** a small fire involving bedding in a berthing compartment was controlled by this system. On *Wasp* (CV18) fires on the second deck were extinguished, using both the fixed fog system and hoses equipped with all-purpose nozzles. On *Ticonderoga* (CV14) a small fire in a space on the second deck to port of the forward elevator was controlled using this system. In these latter three cases, investigation revealed that the fires could have been extinguished just as promptly with all-purpose nozzles and probably with less water. On the contrary, on *Ticonderoga* where other spaces on the second deck were gutted by fire and where the use of the fixed fog system would have been logical, the bulkhead hose connection on the second deck and the hangar deck hose connection could not be reached because of fire and other damage in the vicinity of these connections. On *Kadashan Bay* (CVE76) a fire of burning gasoline from a suicide plane crash, bedding and personal gear in two living spaces on the third deck directly over gasoline stowage tanks and adjacent to gasoline pump rooms was effectively combatted by turning on the sprinkling system. The compartments were then entered with fire hoses using all-purpose nozzles. In this case the fixed fog systems were permanently connected to the firemain.

---

* BuShips War Damage Report No. 16

** BuShips War Damage Report No. 39

*** BuShips War Damage Report No. 47

--28--

4-42 References (q) and (r) outlined the Bureau's policy in regard to the installation of fixed fog nozzle systems. They provided for non-automatic systems for all spaces on the forecastle deck level and below, except for spaces already provided with sprinkling, steam smothering, CO_2 or inert gas protection and for cold storage spaces, tanks and voids. On the basis of war experience and in the interest of reducing weight and conserving labor and material, the policy in regard to the installation of fixed fog systems has been changed as per references (s) and (t). These changes provide that the installation of non-automatic fixed fog nozzle installations be limited to watertight compartments, excepting those indicated in the above paragraph, as follows:

1. Compartments containing gasoline piping and compartments adjacent to these compartments.

2. Compartments containing ventilation ducts which connect with gasoline pump rooms and compartments adjacent thereto.

3. Compartments, exclusive of tanks and voids, which are adjacent to voids surrounding gasoline tanks.

This policy is applicable to vessels now building and those in service where the systems have not been installed. Shipalt CV836 of 7 August 1945 authorizes the removal on ships in service of that portion of existing fixed fog nozzle installations which exceeds the requirements stated above.

4-43 Reports have come from a few carriers indicating difficulty in securing hangar sprinkling and water curtain systems because of damage to control wiring for the electric valve control motors at the hangar deck conflagration station and the hangar deck local station. In such instances the valves reopened after being closed manually and continued to do so until the control circuits were secured at the distribution board since there was no provision for local electrical control at the third deck stations. This has resulted in considerable additional discharge of water in hangars after fires were extinguished. ShipAlt CV510 of 7 May 1945, provides for necessary changes in the electrical controls to enable complete electrical control at the third deck stations.

E. Engineering Notes

4-44 Despite heavy damage above the hangar deck level, *Franklin* did not sustain any direct structural and machinery damage in main or auxiliary machinery spaces. This has generally been true of other CV9 Class carriers which have been damaged by suicide planes and bombs.* It is attributable primarily to the effectiveness of the armored hangar deck (2½-inch STS) and to a much lesser extent to the armored fourth deck (1½-inch STS) which forms the overhead of the machinery spaces. In the action of 19 March, although four fairly large holes were blown in the armored section of the hangar deck as stated in paragraph 3-39, fragments penetrated to the fourth deck in only two isolated places in way of main machinery spaces. Fragments from the bomb, presumed to be a Tiny Tim rocket, which detonated while lying flat on the hangar deck in way of the armored after bulkhead of the forward uptake space (frame 100), traveled down the air intake duct to the fourth deck. The deck was pitted but not penetrated. Fragments from another bomb also presumed to be a Tiny Tim rocket, which detonated on the starboard side of the hangar at about frame 146, penetrated to the fourth deck but caused only a minor depression in the deck.

---

* For example, in *Wasp* (CV18) action of 19 March 1945 a 250 kg. SAP bomb penetrated armored flight deck and detonated at third deck level, frame 128, over No. 4 fireroom. Direct damage was limited to bulging of armored fourth deck. Considerable indirect damage

resulted in No. 4 fire-room from water and blast traveling down ruptured uptakes and vent intakes.

--29--

4-45 In both the actions of 30 October and 19 March, flooding occurred in firerooms as a result of damage to boiler air intakes and uptakes in the hangar. In the first action, firefighting water from the hangar drained into No. 4 fireroom through ruptured air intakes for boilers Nos. 7 and 5 and also through the after bomb elevator trunk, into the boiler air intake plenum chamber (B-431-E) on the fourth deck where it overflowed the coamings around the individual air intake openings. Flooding was finally controlled by the fireroom bilge eductors, but only after fires in boilers Nos. 7 and 8 had been extinguished. These two boilers were not damaged by the flooding but remained out of commission until they could be dried out slowly and fired again. Circumstances were somewhat similar in the action of 19 March although the final damage was more extensive. In this action the STS bulkheads of the forward uptake space in the hangar were ruptured by Tiny Tim explosions at frames 93 and 100 and the boiler uptakes and air intakes within were ruptured at the hangar deck level. Firefighting water drained from the hangar deck through the air intakes to the plenum chambers for the forward boilers on the fourth deck (B-411-E and B-415-E) where it overflowed coamings around the individual air intakes and then drained into boilers Nos. 1,2,3 and 4 in Nos. 1 and 2 firerooms. Some water also drained down the uptakes causing a flare-back in boilers Nos. 3 and 4. Firing of these boilers was continued until water in the fireboxes actually rose high enough to extinguish fires. Firefighting water also drained into boilers in the No. 4 fireroom, but not in sufficient quantity to cause operating difficulties. This water traveled from the hangar via the damaged air intake duct at frame 130 to plenum chamber B-431-E and thence to the fireroom, as in the action of 30 October. High water marks on the bulkheads of B-431-E showed that a small amount of water had overflowed the coamings around the individual air intake openings. Fortunately, the intake duct was not seriously ruptured at the hangar deck level for had this been the case the No. 4 fireroom probably would have been put out of commission also.

4-46 In the two forward firerooms superheat was maintained at 850 degrees F. until the boiler fires were extinguished by flooding. This resulted in a fairly rapid quenching action on lower pressure parts, but a subsequent fireside inspection and a 600 psi hydrostatic test of pressure parts of all forward boilers did not reveal any leaks or temperature cracks in lower drums and superheater headers. The metallurgical characteristics of pressure parts used in modern boilers are such that increased stresses due to sudden application of water at sea temperature will not cause failure. Reference (i) comments on the small amount of apparent damage that was sustained by refractory material due to quenching. No visible damage occurred to furnace brick or insulation. The chrome ore baffles on the saturated side of fire screen tubes washed away to some extent, exposing tubes and studs to the fire in spots. It was noted that contact of the water with pressure parts removed most of the slag and soot. After the forward firerooms were evacuated because of heat and smoke, water continued to drain into the boilers. High water marks in the furnaces indicated that water rose to a depth of 7 feet in the saturated furnaces and to about 6 feet in the superheat furnaces of boilers Nos. 1 and 3 (starboard boilers) and to a depth of about 4 feet in all furnaces of boilers Nos. 2 and 4 (port boilers). The difference in water levels of port and starboard boilers was due to the starboard list during the period that water was draining from the hangar. When these spaces were re-entered, all boiler furnaces were drained and dried out slowly. The fireside of all pressure parts was washed out with fresh water and sprayed with consol oil. After being thoroughly dried out and hydrostatically tested, boiler No. 1 was lighted off and used on the return voyage to the United States. It is believed that the other forward boilers could have been put into operating condition by repairing the uptakes if circumstances had so required.

4-47 In order to decrease the possibility of vent ducts and bomb elevator trunks being torn off flush with hangar deck plating and thus permitting water from firefighting to drain to spaces below, the Bureau

--30--

authorized installation of 3/4-inch STS coamings, 12 inches high, around the bottom of all such ducts and trunks in the hangar as per reference (1). As an additional measure to prevent water which might enter the boiler intake plenum chambers (B-431-E, B-423-E, B-415-E and B-411-E) on the fourth deck from overflowing the individual boiler air intakes, the Bureau authorized extending present 24-inch coamings to a height of 36 inches and installing 6-inch drains, port and starboard in the floodable portion of each plenum chamber as per reference (u). The drains from each space are to be led to the bilge of the fireroom below and are to be fitted with gooseneck traps and proper strainers. Improved drainage facilities in the hangar, which have been authorized, will also reduce the hazard of water from the hangar draining into spaces below (See paragraph 4-95).

4-48 Smoke and heat, which almost invariably enter machinery spaces via ventilation supply systems when there is a fire of any appreciable size in the hangar or on the flight deck, have presented a serious hazard and a difficult problem to combat. Since ventilation must be supplied continuously to machinery spaces, even during periods of General Quarters, it is almost inevitable that some smoke and heat will be drawn in. This occurred in *Franklin* on 30 October and 19 March and on every carrier which has sustained a flight or hangar deck fire of serious proportions. As a matter of fact, this hazard has not been limited to carriers alone for battleships, cruisers, destroyers and other classes of ships have experienced similar difficulties. On *Princeton* (CVL23), for example, dense smoke forced all machinery spaces to be evacuated within 20 minutes after the fire started and the fact that all power and firemain pressure was lost was definitely a contributing factor in the loss of that ship. Individual ships have improvised various means to keep machinery spaces manned during periods of firefighting, but major credit is due to the perseverance and willingness of engineering personnel to endure almost unbearable heat and smoke. In the action of 30 October, *Franklin's* engineering spaces quickly filled with dense smoke. Supply blowers were secured and exhaust blowers were left running. Some engineers donned gas masks which were partially effective, and some wrapped wet towels around their faces while others endured conditions without protection. A few men were provided with rescue breathers which were satisfactory but these required initiating in fresh atmosphere and were inadequate in number. Heat was terrific and although watches were changed often, many men were overcome. When the course of the ship was changed putting the smoke on the port side, supply blowers to enginerooms on the starboard side were operated which greatly alleviated conditions in these spaces. In the action of 19 March, conditions were similar only much more severe. The fire was so intense and widespread that it was not possible to operate any supply blowers to engineering spaces without drawing in heat and dense smoke. Visibility was almost zero with all lights on and even sealed beam lights were of little value. Heat and smoke became progressively worse until it was unbearable and the spaces were then evacuated. Changing watches was handicapped because of the difficulty of egress and ingress through smoke-filled spaces on the second, third and fourth decks. Although ventilation supply to the forward Diesel fire pump space (A-605-AE) was secured, smoke entered through the storage battery vent and forced evacuation of this space.

4-49 It has been reported by several carriers that smoke has been successfully cleared from machinery spaces by securing all ventilation supply blowers and opening access hatches to the third deck. With the exhaust blowers operating, air will be drawn from the third deck which in turn is opened up to forward compartments. This plan might operate successfully if the third deck is clear of smoke but in view of the fact that it involves opening up the third deck (the damage control

--31--

deck), which obviously destroys vital watertight integrity, it should be employed only where conditions permit and where absolutely necessary in order to keep the main plant in operation.

4-50 On the earlier CV9 Class carriers the ventilation systems for the machinery spaces as originally designed and installed had a serious defect in that the port longitudinal vent trunk on the second deck, which takes its supply through openings on the port and starboard quarters of the hangar, was the sole source of supply ventilation for all firerooms and the after auxiliary machinery space. The two enginerooms and the forward auxiliary machinery space were also supplied with ventilation from this port longitudinal vent duct, but in addition were each provided with individual supply ventilation systems taking air from starboard.

4-51 When *Lexington* (CV16) was torpedoed in December 1943, shock of the detonation initiated action of smoke generators on the fantail which released large quantities of FS smoke in the vicinity of the port and starboard intakes for the port vent trunk. Smoke was carried forward and drawn into all spaces served by this trunk, causing many personnel casualties and forcing operating personnel to don gas masks and rescue breathers in order to continue at their stations. Since then there have been 9 additional cases of battle damage to CV9 Class carriers in which smoke in engineering spaces was a major problem. In all these cases the port vent trunk was the source of smoke into these spaces. Later carriers, beginning with CV33, do not have the port longitudinal vent trunk but instead take ventilation supply from separate vent ducts, port and starboard.

4-52 As a temporary measure ShipAlt CV581 of March 1944, authorized sectionalization of the port vent trunk to provide individual port intakes; however, due to the magnitude of work required, few ships have had the complete job accomplished. A number of ships have received partial sectionalization which provides for forward machinery spaces to take ventilation supply from individual port side ducts and after machinery spaces to continue to take supply from the after section of the vent trunk. This is an improvement over the original installation in that the forward and after firerooms will have separate sources of supply. Sectionalization of the vent trunk as provided by ShipAlt CV581, however, does not completely solve the problem for it simply distributes the risk and diminishes the probability of one hit involving the ventilation supply for all machinery spaces. The four firerooms and the after auxiliary machinery spaces will still have supply ventilation from the port side only. The natural tendency of Commanding Officers in the advent of heavy smoke conditions, is to put the wind on the starboard side in order to keep the island and ship control spaces as clear as possible. This will blanket the port intakes authorized by the above ShipAlt and cause smoke to be drawn into those spaces served by port supply systems only and to those spaces served by both port and starboard supply systems where the port systems are not secured in time. It is significant that in eight of ten cases of severe battle damage smoke was eventually cleared from the two enginerooms and the forward auxiliary machinery space by closing the port supply systems and operating the individual starboard supply systems.

4-53 ShipAlt CV864 of January 1946, authorized for the CV9 Class an auxiliary means of obtaining ventilation for the firerooms from the starboard side of the ship. Basically, this source of air from starboard is obtained by securing the existing inlet ducts between the fireroom supply blowers and the port vent trunk and providing by-pass ducts from the uptake enclosures to the existing supply distribution center in each fireroom. The success of this system depends entirely upon the ability of the engineering personnel aboard ship to maintain an indraft through the existing forced draft air intake ducts by means of judicious use of the forced draft blowers. Dampers are provided in both the new and existing duct work. Supply normally will still be taken from the port vent trunk and only in the case of smoke entering the port intakes will the starboard system be used.

--32--

4-54 In the action of 30 October, *Franklin* reported that rescue breathers were invaluable but that the number (106) furnished was insufficient. Since that time, the allowance has been increased to 500 but unfortunately the total increased allowance had not been received by 19 March and only two sets were provided for each machinery space. These were useful though obviously inadequate in number. Also, the fact that rescue breathers must be initiated in clear atmosphere restricts their use in machinery spaces. At the present time the air-line mask appears to be the most practical solution to the smoke problem in machinery spaces. The Firefighting Manual gives complete details and instructions for improvising hose or air-line masks by using the face piece of a Mark IV gas mask, a length of air hose and a cylinder of air with a pressure regulator. *Randolph* (CV15) reported that a number of these units, which apparently were made up according to instructions in the Firefighting Manual, were used and proved to be very effective in smoke-filled machinery spaces. *Missouri* (BB63) and *Enterprise* (CV6) also reported satisfactory results with improvised air masks. An improved air-line mask has been developed and is being manufactured on a large-scale basis. Deliveries are being made to all aircraft carriers. Complete equipment for 50 watch stations is being furnished CV9 Class carriers. Instructions for installation and use of this equipment are covered in reference (1).

4-55 In the action of 19 March, when it was decided that machinery spaces would have to be evacuated because of heat and smoke, it was manned to set the throttles for 8 knots speed and to leave vital units operating and the boilers steaming. Before these spaces were evacuated, however, all boilers were secured. As noted in paragraph 3-45, the two emergency 250 KW Diesel generators started automatically as designed but unfortunately operated at no load. On the after emergency Diesel generator board, the automatic relay device operated to disconnect all loads except the steering gear and failed to reconnect them when the steering gear starting load decreased. Even if it had reconnected the designated emergency loads it is probable that, because of short circuits in damaged areas, the generator circuit breaker would have operated to trip all loads. This occurred on the forward emergency Diesel generator board. In the confusion of heat and smoke, the boards apparently were not stripped and circuit breakers were not closed until the spaces were remanned, approximately six hours later. If this had been done prior to evacuating machinery spaces, power to a few vital circuits from these emergency generators would have been available and possibly would have enabled personnel to reman machinery spaces sooner.

4-56 The three 60 KW casualty power Diesel generators located in the hangar could not be utilized because of the fire in this space. A study of the available electrical energy sources disclosed that these facilities were adequately dispersed and that ample emergency power was available without the use of these three 60 KW generators. Therefore, ShipAlt CV837 of 27 December 1945, authorized the removal of all three generators as partial weight compensation for the high capacity fog foam installation.

4-57 Modification of the casualty power system to incorporate the latest design features and to provide maximum reliability plus greater flexibility of casualty power connections under battle damage conditions has been authorized by reference (v) and various preceding ShipAlts. The most important modifications are as follows:

1. Additional risers are to be installed so as to provide two risers from each space containing either a ship's service or emergency switchboard. A circuit breaker and casualty power terminal are to be installed on each switchboard with a connection between the generator circuit breaker and disconnect switch (ShipAlt CV872).

--33--

1. Risers to be increased in size from 20,000 to 60,000 CM: (ShipAlt CV606).
2. Selected power panels are to be provided with casualty power terminals (ShipAlt CV611).
3. Casualty power bulkhead terminals to be installed between all machinery spaces (ShipAlt CV616).
4. All casualty power risers are to be terminated on the second deck and bulkhead terminals, portable cables, etc. are to be moved from the third to the second deck. The second deck is less likely to be flooded out than the lower decks. (ShipAlt CV871).
5. Casualty power terminals with watertight enclosure are to be installed for each lighting transformer except those located in the same space with ship's service or emergency switchboards (ShipAlt CV873).
6. Sufficient portable casualty power cable to be provided to make connections from all casualty power loads to the distribution system, plus 25 per cent additional (not to exceed 600 feet).

F. Gasoline System and Plane Defueling

4-58 In the action of 19 March, it was reported that gasoline continued to flow from the system in the after end of the hangar for some time after the initial damage and that gasoline also poured out of the high overboard discharge and from outboard filling and drain lines which were fractured by fragments. At the time of the initial damage, the forward gasoline system was secured and the mains purged with inert gas. The after system, however, was in operation. Topping off of planes on the after end of the flight deck had just been completed and three planes were being topped off in the after end of the hangar. Men manning the filling lines were killed at their stations. The man on watch in the after gasoline pump room immediately stopped the pumps and secured the system so as to permit gasoline in the lines to drain back into the stowage tanks. He then left the pump room, closed and dogged the hatch on the second deck, and pulled the CO_2 control. The trunk and pump room subsequently flooded to a depth of about 20 feet through the hatch at the second deck which apparently had not been tightly dogged. The motor room flooded from the pump room through a shaft gland.

4-59 At the time of the hits the after gasoline lines were probably filled to the level of the gassing stations on the flight deck. Under normal conditions, without the eductors in the salt water displacement system being in operation, the process of draining the gasoline back through the system to the stowage tanks is slow. The gassing hose lines and the piping systems on the flight deck and in the hangar were damaged by blast and fragments and whatever gasoline was present in the mains above the hangar deck level no doubt drained into the hangar. The amount of this gasoline, however, at most could not exceed a few hundred gallons and its effect on the fires was insignificant when compared with the nearly 25,000 gallons contained in aircraft fuel tanks on the flight deck and in the hangar. This large amount of fuel is sufficient to sustain a severe fire for several hours. It should be noted that these fuel tanks did not all immediately rupture and explode but instead ruptured singly over a period of several hours as a result of succeeding bomb explosions and the roasting effect of the spreading fires. This of course tended to prolong the duration of the gasoline-fed fires. Burning gasoline floating on the surface of firefighting water at times poured out starboard hangar deck openings and burned on portions of the shell and adjacent sea areas (Photos 15, 16 and 17).

--34--

4-60 The salt water supply to the gravity tank of the gasoline system had not been secured and if the internal float valve which regulates the water level in the tank became stuck, as the ship reports had happened previously, the gravity tank would completely fill. Salt water, not gasoline as reported by some observers, would then be forced out of the high overboard discharge at the second deck as well as at the low discharge at the fourth deck.

4-61 In general, the gasoline system performed satisfactorily. The fact that the after system was full of gasoline to the height of the gassing stations on the flight deck and planes actually were being gassed at the time contributed to the intensity of the fires by only a negligible amount in this case. Even had the system been secured and purged the protection afforded by the inert gas would have been lost by broken lines, it should be noted that the inert gas protection system was developed for use on carriers as an everyday safety precaution measure and has no real damage control value.

4-62 It was two days before dewatering of the after gasoline trunk and pump room could be started because the trunk terminates in 2-207-1 EL which together with adjacent spaces on the second deck was flooded to a depth of about 5 feet. When these spaces were dewatered the gasoline trunk and pump room were pumped out with submersible pumps without difficulty. The ship's force was reluctant to use an electric pump to dewater the spaces, due to the possible presence of gasoline fumes. It was suggested that an eductor operated from the firemain be installed permanently in this trunk. However, portable eductors are available for use with fire hose and are considered sufficient for the purpose.

4-63 Gasoline in aircraft fuel tanks on the flight and hangar decks has been the major factor in the duration and severity of fires on carriers. Doctrine on most carriers requires that all planes be defueled when their missions have been completed and they are returned to the hangar. This was the general practice on *Franklin* when a strike was not imminent. Other ships follow a rule of defueling all planes except fighters. War experience, particularly since the advent of suicide plane attacks, has definitely emphasized the importance of defueling all planes when stowed in the hangar. With present facilities, defueling of planes is a comparatively long and hazardous process. For example, one CV9 Class carrier reported that to defuel one plane presumably a torpedo bomber fully gassed and with auxiliary tanks) required a minimum of one hour with a trained crew. While in most instances planes to be defueled will have completed flights and consequently will not have full tanks (thus requiring a shorter defueling time), it is obvious that present facilities for this task are not adequate. The ideal maximum time required to defuel a plane should approach that required to fuel the same plane. Present air-driven rotary defueling pumps have a rated capacity of 30 gpm which theoretically should defuel a fully gassed TBM-3 (810 gallons) in 27 minutes. Defueling pumps were originally designed to be portable, but at the request of Forces Afloat, were installed in a fixed position adjacent to each fueling station. This presumably was due to the crowded conditions with a full load of planes parked in the hangar deck. The performance of the pumps when permanently located, however, was found to be extremely poor due to the high suction lift inherent in the use of fixed units with long hose lengths plus the high volatility of aircraft gasoline. Accordingly, the Bureau authorized by reference (w) the return to the use of portable pumping units on the flight and hangar decks.

4-64 Vapor will be formed in a gasoline defueling system under any condition which causes the pressure to be reduced to the point where light hydro-carbons in the gasoline are released as vapor.

--35--

This critical pressure is difficult to determine on shipboard since it occurs at various points in the system and the operating personnel are not aware of its presence until the pump loses suction. Presence of vapor necessitates venting the entire system which largely curtails defueling operations and is a tedious and time-consuming process. Investigation by the Bureau has shown that the current practice in the use of defueling pumps should be entirely revised and that minor modifications to the airplane defueling connections are required in order to reduce the hazard of vapor-locked gasoline in the suction side of the defueling lines and to increase the efficiency of the pumps. A new portable defueling unit has been developed using a small oil-filled, explosion-proof motor driving a centrifugal pump. This unit is rated at 60 gpm at a 35-foot head and weighs only 28 pounds. A suction manifold with multiple hose connections will be used. This unit, with a discharge hose, can be easily handled by a crew of two men and is sufficiently small and light to permit its being carried direct to the airplane where it can be connected to defueling connections through only 4 to 5 feet of 1¼-inch suction hose. Thus, practically the entire system is on the discharge side of the pump and subject to pressure, minimizing the release of vapor. A supply of these pumps has been procured and tests are now in process on board a CVB. Pending results of these tests, new carriers have been furnished with portable air-driven pumps for the hangar deck stations only. In addition, the Bureau of Aeronautics is currently investigating the need for, and feasibility of providing individual 1¼-inch drain connections at the low point of each aircraft fuel tank, with the shortest possible piping leads, which will not only permit complete removal of gasoline but will put a positive head on the suction side of the pump.

4-65 It has also been reported by some carriers that the time required to drain gasoline-back to stowage tanks and to purge the system with inert gas was excessive. In one instance it was reported to be approximately one hour, which obviously is unacceptable. It is presumed those ships which have experienced this difficulty were not provided with the proper eductors in the salt water displacement system as authorized by reference (x). This installation should also be helpful in expediting transfer of gasoline from tankers. Tests have shown that carrier gasoline systems which are properly equipped and operated can be secured and purged in an average time of five to ten minutes.

G. Flooding and Stability

4-66 CV9 Class carriers tend to list to starboard when firefighting water is shipped below decks in large quantities. *Franklin* developed starboard lists in both the actions of 30 October and 19 March. This has been erroneously reported as resulting from off-center flooding of the after elevator pit. While the after elevator platform (including the auxiliary elevator platform) is somewhat to starboard, the elevator pit proper is on the centerline and flooding will not result in off-center moment. In all cases of considerable flooding below decks, water has drained down from the hangar deck via starboard side sources, which include open access hatches, damaged vent and air intake ducts, uptakes and bomb elevator trunks. Because of non-watertight bulkheads on the second deck between frames 79 and 131, water has been free to flood all spaces between these frames from the inboard bulkhead of the port vent trunk to the starboard shell. These compartments on the second deck do not have to be watertight to meet stability and floodable length requirements. Although war damage experience has shown advantages in favor of additional watertight bulkheads in these spaces, this alteration has not been undertaken due to the excessive amount of ventilation and electrical interferences in the overhead. Longitudinal bulkheads of the port vent trunk are watertight and have prevented spaces to port of the trunk from flooding via sources

--36--

enumerated above. Third deck spaces have flooded through non-watertight hatchless companionways on the second deck in the area between frames 79 and 131. In addition, all decks are subject to direct flooding to the extent structural bomb damage penetrates to these decks, as occurred on *Franklin* in the action of 19 March.

4-67 In the action of 30 October the ship assumed a starboard list which gradually increased to a maximum of 3 degrees. At this point counterflooding measures were started and with flooding of 11 port damage control voids the ship slowly returned to an even keel and then gradually assumed a port list of 2 degrees. Port compartments between frames 79-142 on the second deck and decks below outboard of the port vent trunk gradually flooded. The shifting of water together with the counterflooding measures produced a port transverse moment more than sufficient to offset the initial starboard moment and explains the resultant port list. It is estimated that approximately 2150 tons of water accumulated on the hangar deck and below. The estimated displacement before damage at the reported mean draft of 28 feet was 35,520 tons. The corresponding GM at this displacement was 9.55 feet. Estimated GM after damage as calculated by the Bureau for the worst possible condition of free surface and off-center moment was 3.7 feet at a displacement of 37,670 tons.

4-68 In the action of 19 March, water in large quantities drained to spaces below the hangar deck via the damaged starboard air intakes, uptakes, open access hatches and also through the large bomb holes blown in both the armored and unarmored portions of the hangar deck. Flooding of second deck compartments inboard of the port vent trunk resulted in an initial starboard list and pocketed free surface water aggravated this condition. By 0815, slightly more than one hour after the initial damage, the reported list was 3 degrees and by 0954 it had increased gradually to 8 degrees. From then on until the maximum starboard list of 15 degrees was assumed, the increase was estimated to be at the rate of about 1 degree every 10 minutes. The volume of water being discharged by *Franklin's* firefighting systems was augmented appreciably by hose streams from other ships, all of which were directed into the starboard side because of the difficulty of approaching the port side in the face of heat and smoke.

4-69 An accurate analysis of the effect of this free water on the stability characteristics is impossible since no definite data is available concerning the amount of water shipped below and its chronological appearance in various compartments. It is estimated, however, that not more than 5000 tons of water in total accumulated below the hangar deck. Based on the worst possible flooding conditions, by assuming simultaneous flooding (with free surface) of all centerline and starboard compartments which inspection revealed to have been partially or completely flooded, and by assuming the maximum likely flooding of hangar and gallery spaces, calculations show that the GM when the ship was listed 15 degrees was about 2.5 feet with a displacement of about 44,000 tons. From this point on, the GM improved slowly as various compartment's flooded completely, thus removing free surface effect, and as the water in the gallery and hangar spaces drained away. Counterflooding of about 23 port voids plus slow flooding of port spaces on the second deck outboard of the longitudinal vent trunk resulted in the ship gradually coming upright (at 0000) and then gradually listing to port, coming to rest at about 9 degrees at 0400. The GM at this time is calculated to have been about 6.5 feet. En route to Ulithi, certain of the port voids and compartments were gradually dewatered and the ship slowly returned to even keel.

--37--

H. Damage Control Notes

4-70 Some of the items presented in this section are primarily operational problems and are not normally discussed in these War Damage Reports. Such matters are mentioned hereinafter in order to disseminate information of probable future value.

4-71 *Franklin* is an outstanding example of improvements accomplished since the early part of the war in the design and construction of ships to resist damage and in damage control equipment, technique and training of personnel. Damage Control and Firefighters' Schools and the program of education in firefighting and dissemination of knowledge of damage control through the medium of FTP-170-B and other publications have definitely proved their value. Training of the entire ship's company as well as damage control personnel in the technique of damage control, particularly in firefighting, has paid excellent dividends. *Franklin's* experiences have again demonstrated that damage control is an all-hands problem and that personnel regularly assigned to that function must be prepared insofar as possible to perform efficiently in all parts of the ship as well as their assigned stations. For example, in the action of 30 October, 18 men of Repair I were trapped in the shipfitters' shop and until they could be released about one hour later the hangar deck repair party required augmentation from other repair parties. Again, in the action of 19 March, many repair party personnel were killed by the initial blast and others were trapped or blocked by fire, smoke and damaged structures.

4-72 Valuable lessons in damage control technique have emanated from *Franklin's* experiences and these are being disseminated to the Forces Afloat and to various Damage Control and Firefighters' Schools. Many ships in advance areas made a practice of sending their damage control officers to inspect damaged ships and interview personnel in order to gain first-hand information. For example, after *Franklin* was damaged on 30 October and put into Ulithi, *Intrepid* sent a party to inspect her. On the basis of this inspection and interviews with ship's officers, the damage control officer prepared a memorandum which is quoted in part (with Bureau comments) below:

1. "The *Franklin* Hangar Deck Control Station was knocked out by the effects of blast; hence the Central Sprinkler and water curtain station could not function and control had to be assumed manually at the local control stations.

   "RECOMMENDATION: Local control booth stations should be manned by Repair I personnel, as outlined in enclosure (A). This is not at present possible with the limited number of men assigned Repair One."

Bureau Comment: This occurred again in the action of 19 March and also on other carriers. Armoring of the hangar deck control (conflagration) station and wiring, which has been authorized as noted in paragraph 4-33, will provide better protection. Manning of all local control booths and manual valves for controlling hangar sprinkling and water curtains is an operational matter and appears to be desirable.

1. "Repair I personnel on the *Franklin* number 55. This number was insufficient to fight the fire.

   "RECOMMENDATION: Inasmuch as the present number of men assigned Repair I on this ship is much less than 55, it is recommended that the number now assigned be supplemented at General Quarters and at Fire Quarters by Aviation Shop personnel as indicated in enclosure (A)."

Bureau Comment: This is an operational matter.

--38--

1. "Personnel in Ready Rooms, on Flight Deck and on Island Structure of the *Franklin* were endangered and in many cases injured or killed as a result of the plane crash and bomb explosion.

"**RECOMMENDATION:** That all pilots and aircrewmen lay down to the second deck or below at General Quarters unless a strike is imminent and their services may be required. No unauthorized personnel should be in the Ready Rooms, on the Flight Deck or on the Island structure during General Quarters."

Bureau Comment: This procedure was being followed generally by *Franklin* at the time of the damage on 19 March and saved many air personnel. Nos. 2, 3 and 4 Ready Rooms will be relocated to the second deck. This matter is discussed more fully in paragraph 4-86.

1. "Personnel in Shipfitter Shop were trapped therein with no means of escape.

   "**RECOMMENDATION:** Ladder should be secured outside of Shipfitter Shop to Flight Deck. This is particularly necessary since Repair I personnel find shelter from the general exposure of the Hangar Deck within this shop. NOTE: This is being accomplished by the ship's force."

Bureau Comment: During the availability following the damage of 30 October a ladder from outboard of the shipfitters' shop to the flight deck was installed on *Franklin*. In the action of 19 March this ladder served its designed purpose in permitting repair party No. I personnel to escape via this ladder to the flight deck. In an action subsequent to 30 October, *Intrepid's* Repair I personnel also used the ladder to evacuate the shipfitter shop. This ladder plus many additional ladders have been authorized for the CV9 Class.

1. The *Franklin* developed a starboard list as a result of the great amount of water which collected on the Hangar Deck and in the starboard compartments immediately below. The accumulation of water with its surface skin of gasoline resulted in the spread of the fire throughout the 2nd, 3rd and Hangar Decks, endangering the ship.

   "**RECOMMENDATION**: In the event of a large fire on the Flight or Hangar Decks, it is recommended that the Bridge immediately develop a port list until Damage Control can take necessary measures to create a temporary list to port by movement of liquids - thus to flow water and gasoline off of Hangar Deck out of opened port side."

Bureau Comment: In each of the two actions in which *Intrepid* sustained fires in the hangar as a result of suicide plane crashes, a temporary list to port was produced by executing starboard turns until a permanent list could be obtained by flooding port voids. This proved beneficial in freeing the hangar of firefighting water and burning gasoline. As noted in paragraphs 3-21 and 3-47 starboard lists developed on *Franklin* in both actions before measures could be effected to put a port list on the ship. It appears that if the ship had been intentionally listed to port at an early stage in the action, the ultimate flooding would have been less. With additional hangar deck drainage facilities as authorized by reference (u), this procedure will be even more effective in discharging firefighting water and free gasoline over the side and would lessen the intensity of the fire in way of the island structure.

--39--

1. Gasoline flowed down hatches to second and third decks.

   "RECOMMENDATION: That a 2" half round coaming be installed around Hangar Deck hatches such that it will stop the flow of liquids to decks below but will not interfere with aircraft operations. This will be accomplished by the ship's force."

Bureau Comment: Reference (y) authorized installation of 8-inch coamings around hanger deck hatches.

1. Hose lines were of insufficient length to reach fire area and there was a shortage of hose with which to extend these lines.

   "RECOMMENDATION: The hose supply is adequate throughout this ship."

Bureau Comment: Reference (m) authorized additional hose at fireplugs in the flight deck, gallery walkways and in the hangar.

1. Armed machine guns on Hangar Deck planes created a considerable hazard in Hangar Spaces.

   "RECOMMENDATION: When possible that ammunition be taken out of planes at the end of strikes. In this connection it is further recommended that no bombs be stowed on Hangar Deck except immediately before loading on planes. If there had been bombs in planes on *Franklin's* Hangar Deck without a doubt the ship would have been lost."

Bureau Comment: The exploding of small caliber ammunition presents but little actual hazard to personnel and is not particularly destructive to material. As demonstrated on *Franklin* in the action of 19 March, bombs in the hangar and on the flight deck constitute a most serious hazard. Insofar as operating circumstances permit, bombs, rockets and torpedoes should be kept in the magazines.

1. "Stowage of planes on the Hangar Deck made no allowance for movement of Damage Control personnel or equipment and thereby impeded the progress of personnel engaged in fighting the fire.

   "RECOMMENDATION: That the present spotting of planes on the Hangar Deck be studied and that, when practicable, an aisle of about 2 feet be created fore and aft to allow movement of personnel, also that the plane customarily spotted in close proximity to the Main Locker of Repair I be so placed as to allow Repair I personnel access to the Hangar Deck in any direction at all times."

Bureau Comment: Although an operational matter, this is very important.

1. "Firefighting efforts of personnel were hindered by lack of adequate ventilation in smoky compartments.

   "RECOMMENDATION: That Supply Department furnish Damage Control with information as to the location of its portable blowers so that, in an emergency, such blowers may be utilized in dispelling smoke."

Bureau Comment: This is an operational matter. Improvements in ventilation supply systems have been authorized and will alleviate some of the smoke difficulties.

--40--

1. "Flash proof covers were not on bunks. This resulted in a great deal of very nauseous smoke and a very persistent smoldering fire in several of the living spaces.

   "RECOMMENDATION: That when in a battle area, personnel be required to keep flash proof covers on and under mattresses when bunks are not occupied; that personnel be cautioned when leaving bunks for battle stations to take time out to envelop bunk with flash proof cover."

Bureau Comment: Flashproof covers have proved their worth and should be kept on all bunks including those in officers' country.

1. "Gasoline from plane found its way down bomb elevator to Third Deck - causing explosion and spread of fire.

   "RECOMMENDATION: That flash covers on bomb elevator doors be kept closed at all times when bomb elevators are not in use. The existence of this condition to be checked by cognizant personnel at General Quarters."

Bureau Comment: Doors at the fourth deck level of bomb elevator trunks are being replaced by dogged hatches per ShipAlt CV780 of 10 April 1945. The installation of STS coamings on vent trunks and bomb elevators penetrating the hangar as per reference (1) will reduce this hazard.

1. "Repair party personnel were apparently not sufficiently versed in the use of Rescue Breathing Apparatus.

   "RECOMMENDATION: That smoke bombs be used in the after incinerator to permit a Rescue Breathing Apparatus drill under actual smoke conditions."

Bureau Comment: This is an operational problem.

1. "Repair Party personnel were unable to function in many cases because of a shortage of Rescue Breathing Apparatus.

   "RECOMMENDATION: Increase in allowance of Rescue Breather Apparatus by 50. Letter stating necessity for increase is being prepared."

Bureau Comment: Allowance of Rescue Breathers has been increased to 500 for the CV9 Class.

4-73 The information gained and lessons learned from the inspection of *Franklin's* damage were effectively employed by *Intrepid* when she was damaged by two suicide planes on 25 November 1944, with a resulting conflagration very similar to that of *Franklin* on 30 October. As a matter of interest, *Intrepid* was extensively damaged a second time on 16 April 1945, and again the damage, including a severe fire in the hangar, was handled in an excellent manner.

4-74 The importance of holding frequent realistic drills in damage control cannot be overemphasized. Few ships actually operate hangar sprinklers and water curtains before damage, yet there is no other way to properly acquaint all hands with the function of the many control valves and the problems of draining off the large volume of water which deluges the hangar deck. A good example of this point is noted in the War Damage Report of *Wasp* (CV18) for damage sustained when hit by a 250 kg SAP bomb on 19 March 1945. This is quoted in part below:

--41--

"While in port, all aircraft are moved to the flight deck and the hangar curtains actually turned on; control being exercised by both the conflagration control station and at individual control stations.... Repair party personnel have been required to know not only their own individual job, but all jobs on the team, perfecting this knowledge by repeated practice. These policies resulted in *Wasp* being able to maintain station at all times.... The fire in the hangar deck was rapidly isolated and extinguished."

4-75 The attack of 30 October caught *Franklin* in Material Condition BAKER with her entire complement of planes on board, most of which were fully gassed. Although setting of Condition ABLE had been ordered, complete closure had not been effected when the hit occurred; the fact that one armored hatch in the hangar deck remained open contributed appreciably to damage below decks. It is significant that the procedure on *Franklin* at that time required separate orders for General Quarters and setting Condition ABLE. For example, at 1419 General Quarters was sounded and at 1423 setting of Condition ABLE was ordered. At 1426, seven minutes after General Quarters was sounded and three minutes after setting of Condition ABLE was ordered, the enemy plane crashed through the flight deck. The man assigned to close the hangar deck hatch at frame 109, starboard, was killed at his station before completing his assignment.. It is presumed that he was delayed in closing this hatch by personnel going to their battle stations. Burning gasoline and water poured through this hatch to second and third deck compartments. At the time of the attack of 19 March, the Task Group Commander had ordered Condition YOKE set in all ships. On *Franklin*, Condition YOKE was not set, but instead, a modified Condition ZEBRA was set which provided for a somewhat more complete state of closure. One designated hatch in the hangar deck was open to permit access to and from the crew's mess hall. In effect, the material condition of the ship at the time of this attack was comparable to that which existed on 30 October although circumstances were considerably different.

4-76 There appears to be no uniformity in the material conditions as employed by various ships of the CV9 Class. In FTP-170-B, large aircraft carriers are listed as three-condition ships, yet from Action and War Damage Reports submitted by various ships of this class which describe setting of Conditions BAKER and ABLE, it is apparent that at least some of them operated as two-condition ships. In this respect it is noted that in reference (e), which is *Franklin's* War Damage Report covering the action of 30 October, Material Conditions BAKER and ABLE were mentioned, whereas in reference (g) describing the action of 19 March, Material Conditions YOKE and ZEBRA were mentioned. Again, it is apparent that there is a wide variation in the adaptation of modified Condition ZEBRA as employed by various carriers of this class. CNO is reviewing the entire problem of material conditions for the various damage control versus access and living condition requirements. It now appears that additional material conditions will be set up for large vessels which will permit more flexibility and will at the same time be more definitive with respect to the present "modified" conditions. FTP-170-B will be revised accordingly.

4-77 Although war damage experience has not indicated the necessity, the Bureau is now studying and will authorize the establishment of a secondary damage control station, with adequate communications, placed as far from the main station as possible in a well protected area. This station will be located on the third deck (damage control deck) of *Franklin* at the present location of either Repair III or Repair IV and will function only in the event the main damage control station is incapacitated.

--42--

4-78 Rescue breathers have proved indispensable in combatting fires where dense smoke, toxic gases, or a deficiency of oxygen exists. The former allowance of 106 for CV9 Class carriers was found to be totally inadequate for major fires and was increased to 500. On 19 March *Franklin* had on board 208 breathers with 8 spare canisters for each unit. A few isolated reports have been received to the effect that rescue breathers at times failed to function properly but on the basis of the large number of reports of satisfactory service experience it is believed that the difficulties were largely due to incorrect usage. Proper indoctrination and repeated drills by all hands in the use of rescue breathers under realistic working conditions is essential, not only to acquaint personnel with their operation but to instill confidence in the apparatus.

4-79 One limitation in the use of the present type rescue breather is that the initial and subsequent canisters must be activated for oxygen production in clear atmosphere which imposes some limitations on its use. The Bureau and the manufacturer are currently working on the development of an improved type breather and canister which can be initiated in a contaminated atmosphere. It is hoped that when this improved apparatus is perfected that its action will be initiated faster and the canister will last for a longer period. As a temporary expedient the Bureau is currently distributing to Forces Afloat a valve installation for the Type A-1 rescue breather which will permit change of canisters in contaminated atmosphere. Upon removal of a canister from the apparatus the valve acts to seal off the air within the mask and reopens upon insertion of the new canister.

4-80 A small edition of the canister Type A-1 oxygen rescue breathing apparatus has recently been developed and is now under manufacture. This breather, which will be designated as the "Type B," is tended primarily for personnel whose duties will not involve appreciable physical exertion since the rate of output of oxygen is considerably less than the standard Type A-1. The weight of this new breather, including canister, is 6 pounds. It is now planned that about 400 Type B breathers will be added to the allowance list of CV9 Class carriers in addition to the 500 Type A-1 breathers and air-line hose masks presently authorized.

4-81 Reports have been received indicating that Navy service gas masks have been used in smoke-filled compartments when rescue breathing apparatus was not available. Some reports have been favorable, while others indicated that gas masks were unsatisfactory for smoke. In *Franklin's* action of 30 October, there were three men stationed in the laundry (B-429-E) when the crash occurred. Two of them donned gas masks and the third placed a wet towel over his face. Only the man with the wet towel survived. The other two men, wearing gas masks, were later found dead from asphyxiation in B-304-L. It is known that additional deaths resulted under similar circumstances, but this case is the most completely identified with respect to the ineffectiveness of this type gas mask when used for long periods in smoke-filled compartments that are at the same time deficient in oxygen. It is probable that gasoline vapors were present in this instance which in itself might have been fatal as this mask is ineffective against petroleum vapors.

4-82 In the action of 19 March, it was reported that of the 479 bodies recovered approximately 35 per cent were believed to have died from asphyxia. Representatives from the Bureau of Medicine and Surgery, the Bureau of Ships, the Army Chemical Warfare

Service and the Naval Research Laboratory made an inspection of the ship to investigate this matter and their findings are included in reference (j). It was noted in this report that casualties in the hangar, gallery and forecastle

--43--

deck spaces were believed principally due to fire and severe heat. Cause of death in the sick bay (third deck) was not ascertained but it is believed to have been some form of respiratory difficulty. Some personnel were seen alive in this area from one to two hours after the initial damage. There was little evidence of heat and smoke and the bodies were unmarked externally. In berthing and other spaces below decks where bodies were recovered there was evidence of heavy smoke. Reference (j) contains some interesting notes and comments on the use of Navy gas masks in smoke. In one instance, it was reported that 10 men in the forward auxiliary machinery room (fourth deck) donned gas masks after dense smoke entered this space through the ventilating system. Before masks were donned, it was noted that the smoke caused coughing and lacrimation. Smoke was so dense that electric lights were invisible. Personnel remained in this space for about two hours. At the end of one hour, new gas masks were put on due to the fact that resistance to breathing had increased because of the collection of smoke particles on the filter. Heat was severe but not unbearable. That the gas masks filtered out the smoke was indicated by the increased resistance to breathing described by personnel. When a new mask was donned resistance was decreased. Unfortunately, no used gas masks were retained for examination and analysis of deposits in filters. *Enterprise* reported that "gas masks were used to advantage in fighting the flight deck fire, affording protection to firefighters in areas where there was plenty of oxygen."

4-83 On the basis of the service experience it is apparent that Navy service gas masks are reasonably effective against smoke. Personnel must be thoroughly acquainted with their limitations, however. The Firefighting Manual contains a statement that this mask may be used for passage through smoke-filled compartments or for entry into such compartments to close valves or to perform some similar task that can be accomplished quickly. It also contains a word of CAUTION which describes the limitations of the canister, particularly emphasizing the fact that it does not generate oxygen and should not be used in air containing less than 16 per cent of oxygen or in air having a heavy concentration of smoke from oil fires, except for very short periods of time. In every case where smoke penetrates the mask a new canister should be provided prior to futher use. It should also be stressed that the mask offers no protection against gasoline fumes which are apt to be released by damage to carriers.

I. Accesses and Escape Routes

4-84 Quoting in part from reference (g) under "Lessons Learned, Conclusions and Recommendations":

"The gallery deck on present CV's is a death trap. It bears the full brunt of an explosion on the flight or hangar decks and catches the flames of almost any fire in these areas. Personnel in gallery deck spaces aboard *Franklin* at the time of attack, with few exceptions, had absolutely no avenue of escape and were doomed to certain death from the moment when the first bomb exploded.

"The following recommendations are made:

A. Eliminate the gallery deck entirely.

B. Locate CIC below decks next to gunnery plot, with an armored trunk leading to it, having outlets on the second and third deck.

C. All other gallery deck spaces should be located immediately under the hangar deck, which could be raised enough to allow for the new deck beneath it.

--44--

D. If the gallery deck is retained, exits should be provided to the flight deck walkway from each space, as well as quick-acting, spring-loaded emergency hatches to the flight deck."

4-85 That gallery deck spaces are death traps is unquestionable. The original design of the CV9 Class provided for comparatively few gallery deck spaces in way of the hangar, and of these, only the squadron ready rooms were occupied by personnel during General Quarters. Subsequent alterations have provided for additional stations which must be manned continuously, notably CIC and air plot. Because of the high location, gallery deck structures have been lightly constructed. In all cases of major fires in hangars, gallery spaces have been gutted or subjected to such extreme heat and smoke that they were untenable (Photo 38). The light construction, particularly expanded metal walkways, has been very susceptible to extensive damage by blast and fragments (Photos 31, 32, 33, 34).

4-86 The elimination of the gallery deck entirely is not considered practicable in that the already complete utilization of compartments below the main deck prohibits the relocation of most gallery deck spaces. The raising of the armored hangar deck to provide a new deck for the relocation of gallery deck spaces assumes the proportions of a new design problem and is not practical as an alteration to vessels in service or advanced state of construction. However, as a result of the action of 19 March, ShipAlt CV914 of 16 April 1946 authorized for CV9 Class carriers in active status the relocation of three ready rooms to the second deck. No. 2 ready room was relocated to the forward wardroom messroom, frames 54-67. No. 3 ready room and air combat intelligence was moved to frames 79-86, displacing the flag, captain's and navigator's offices which in turn were moved to the gallery deck. These latter offices are not manned during General Quarters. No. 4 ready room moved to frames 100-104 with the group commander's ready room, displacing marine berthing.

4-87 In the action of 30 October, gallery spaces were severely damaged and gutted by fire between frames 110-145. Few personnel casualties occurred on the gallery deck on this occasion, for personnel had sufficient time to evacuate in an orderly manner. CIC was rendered untenable due to smoke and heat. In the action of 19 March, gallery spaces were severely damaged and gutted by fire aft of the forward elevator from frames 50-205. Personnel casualties were heavy. Only one man escaped from CIC and air plot. Since CIC and air plot were almost directly over the center of detonation of the enemy bomb which hit forward and were subjected to severe blast and fragment attack, it is presumed that most of the personnel on watch were killed by the initial blast. Others probably were stunned initially and asphyxiated and burned before they could recover. Information is lacking on the escape route taken by the one survivor (officer) from CIC. He somehow made his way to the forward port 20mm gun platform where he was found in a dazed condition. It is presumed that he left CIC by going aft through the open archway into air plot and thence through the inboard door to the centerline passageway. How he traversed from this point to the 20mm gun platform is a mystery, for the gallery deck forward and to port in this area was blown to the overhead and generally demolished. CIC and air plot each have access doors port and starboard which appear to be adequate. The inboard doors, however, are very apt to be blocked by damage to the lightly constructed decks and expanded metal walkways.

--45--

4-88 At the present time a few of the CV9 Class have CIC located in the hold. This location represents the Bureau's opinion as to the most satisfactory space for the safety and assured operation of this important function. The installation at the gallery deck level has been primarily at the instigation of Forces Afloat in order to keep CIC contiguous with air plot. Although CIC and air plot are protected with 3/4-inch STS plating on the deck and overhead, these spaces are still highly vulnerable because of their location. In the action of 19 March about 90 per cent of the fragments were defeated by the STS plating, but the remaining 10 per cent plus the heat and smoke effectively disabled CIC. Lightly constructed structure adjacent to these two compartments was blown to the overhead.

4-89 The 1 officer and 14 stewards who were lost in the Flag and Commanding Officer's quarters were trapped because of inadequate access. There is evidence that a majority of them survived the initial explosions and were subsequently asphyxiated and burned.

4-90 Steps were taken to improve the number of emergency escapes in gallery and certain other spaces on CV9 Class carriers by reference (z). This letter provides for numerous additional quick-acting escape scuttles, doors, ladders and passageways. All dogged doors leading from the gallery deck to gallery walkways and providing access to or from working spaces, living spaces and spaces manned at General Quarters will be replaced by quick-acting doors. However, there is no remedy for the lightweight metal utilized in the present catwalks and passageways in view of the weight limitations imposed by the high location. Arrows painted in fluorescent paint or fluorescent buttons indicating the direction of egress as an aid to personnel groping in smoke and darkness may be installed by the ship's force as found necessary in organizing the ship for damage control.

4-91 Reference (g) also stated that the island was a death trap because of the absence of exits to the starboard side. While it is apparent that additional emergency escapes on the starboard side of the island are needed, it is a matter of record that comparatively few officers and men have been lost in island structures because of inadequate escapes. Actually, in the case of *Franklin*, there was no report of a single fatality in the island as a result of inadequate escapes. Prior to the action of 19 March, the ship's force installed an emergency escape scuttle in the forward bulkhead of Radio One which accounted for the safety of all 14 members of the watch who otherwise might have been trapped. References (l) and (z) authorized additional emergency escapes from the island and reference (aa) additional accesses which will permit travel from the second deck to the island and flight deck without passing through the hangar. This involves installation of new main deck hatches and rendered existing hatches at frames 75, 109 and 126 starboard unnecessary. ShipAlt CV881 of 10 April 1945 authorized the removal of these three hatches and blanking of the deck. When this alteration is accomplished, one of the principal sources of below decks flooding will be eliminated.

4-92 Sheet metal light locks around a number of ladders in the hangar were demolished or collapsed by blast and fragment attack (Photo 6). This resulted in preventing the escape of some personnel from the hangar. In the island structure all such light locks remained intact although they apparently acted as flues to allow smoke and heat to enter upper decks of the island. To prevent blocking accesses, the Bureau has authorized the replacement of sheet metal light-traps around all inclined ladders with fiberglass cloth mounted on suitable framework as per reference (u).

--46--

4-93 In the action of 30 October, the initial bomb blast traveled down the No.3 bomb elevator trunk and blew out the door connecting the bomb elevator machinery space (B-435-T) with compartment 3-431-E. This allowed firefighting water flowing down the trunk to flood through individual boiler air intakes in B-431-E, boilers Nos. 7 and 8 (Photo 8). To reduce the probability of a similar casualty occurring again, ShipAlt CV780 of 10 April 1945 provides for the replacement of the door to the bomb elevator machinery space on all three upper stage hoists with an 18-inch watertight dogged scuttle with the bottom of the scuttle about 3 feet above the deck.

4-94 In the action of 30 October, the fact that the armored hangar deck hatch at frame 109, starboard, was not closed permitted burning gasoline and water to drain to the second deck and subsequently to the third deck through second deck companionways which were not fitted with hatch covers. Similar cases have also occurred on other carriers. To protect the third deck spaces from flooding and fire from above, ShipAlt CV804 of 22 June 1945 authorized the installation of watertight hatch covers with 18-inch quick-acting scuttles on second deck companionways at frames 94, 109, 148 and 152.

J. Miscellaneous Notes

(1) Hangar Deck and Elevator Pit Drainage

4-95 War experience has demonstrated that excessive accumulation of water on the hangar deck and in the elevator pits from fire-fighting systems will add greatly to the amount of water shipped to below deck spaces through open hatches, ruptured bomb elevator trunks, uptakes and air intakes and bomb and fragment holes in the hangar deck. The free surface effect of water spread over a large area of the hangar deck will also cause an appreciable reduction in the initial stability of the ship. That the water accumulates in such large amounts is due to the small size of the hangar deck drainage openings (about 8×3 inches) as originally designed and to the great amount of debris which usually accumulates on the deck as a result of explosions and prolonged fires which tends to clog up these drainage holes. The deck-edge elevator opening on the port side provides satisfactory drainage for the amidships area if the ship is on even keel and trim, but the 12-inch sheer strake which extends above the hangar deck in way of all other openings prevents free flow of water over the starboard side and also from the forward and after sections on the port side. To provide more adequate drainage facilities, reference (u) authorized cutting the sheer strake flush with the hangar deck in way of roller curtains on the starboard side at frames 45-55 and 166-176, also incidentally providing two more plane jettisoning stations. This letter also provided for installing large freeing ports in the STS bulwarks around the two 40mm quad mounts port side, frames 45-60. The sheer strake inboard of these gun tubs should be cut flush with the deck. No feasible method of providing a large drain from the hangar deck on the port side aft was found although it is believed that the improved drainage resulting from the above alterations will prove adequate.

4-96 Difficulty in draining water from elevator pits has been reported by practically every carrier which has experienced a major fire in the hangar. Present drains are inadequate in size and the strainers clog easily. References (1) and (u) authorized installation of two flush deck drains with suitable strainers in each longitudinal bulkhead of the elevator pit.

4-97 Exhaust ventilation systems which terminate in the forward and after elevator pits have caused the flooding of spaces below the second deck by firefighting water which collects in the pits. It has been recommended that these systems be removed from the elevator pits. The systems are being retained in view of the fact that the exhaust assists in dispelling gasoline vapors which might accumulate in the elevator pits. To reduce the hazard of flooding via these exhaust systems, the Bureau authorized the installation of watertight closures in each ventilation duct opening into the elevator pits as per reference (u). The additional closures are to be classified the same as the fans on each system.

4-98 No drainage facilities are now provided for gallery deck spaces. Several ships have reported that upwards of 500 tons of firefighting water have accumulated in these spaces. This water can be removed only by portable pumps and bucket brigades or by cutting holes in decks to let it drain into the hangar. Cutting drain holes in the decks has been advanced as a simple answer to this problem but is not acceptable as a permanent installation since the fume tight integrity of the structure would be destroyed, air conditioning efficiency would be reduced and unsanitary conditions would result. Drain holes would also present operational difficulties to planes and personnel below. Fixed drainage piping systems would involve considerable additional weight high up for which no adequate compensation could be made. Water on these decks has one desirable effect in that it absorbs heat and thereby offers some protection to the structure. No permanent drainage facilities will be provided.

(2) Elevators

4-99 As evidenced by *Franklin's* actions of 30 October and 19 March and by similar experiences of other carriers, the inboard airplane elevator platforms are so located as to be particularly susceptible to blast damage from explosions within the hangar spaces. The blast effect generally forces these platforms up a few feet and upon falling they assume a canted position within the flight deck opening. The operating plungers, guides, cables, etc. for both inboard and deck-edge type elevators seem about equally susceptible to fragmentation damage. Carrier airplane elevators are of necessity lightly constructed in order to obtain high speed service without undue weight. The inboard main elevators on *Franklin*, for example, are capable of making a round trip in 53 seconds with a live load of 28,000 pounds.

4-100 Since the deck-edge type elevators are inherently more efficient from an operational standpoint in that more continuous flight operations can be maintained, and elevators so located are not as susceptible to blast damage and even if damaged will not compromise the launching and landing areas of the flight deck, future carrier designs will perhaps utilize this type only. One disadvantage of the deck-edge elevator type is its vulnerable position as regards storm damage, particularly on carriers with relatively low freeboard.

(3) Lighting for Smoke Penetration

4-101 As noted in paragraph 3-11, the sealed beam portable lanterns were the only lights which were even partially effective in penetrating smoke in smoke-filled compartments. Even with these lights it was possible to see only a few feet ahead, which was inadequate for efficient firefighting.

4-102 The present damage control lantern which employs a 6-volt sealed beam lamp (Type SB-1) has a candlepower of approximately 200,000 when the batteries are fully charged but weighs only 30 pounds.

--48--

This candlepower is produced by operating the lamp at 8 volts, the maximum amount of overvoltage that can be accepted. It appears that the lantern produces about as much candlepower as is practicable at present in a portable battery-powered lantern of comparable size and weight. Its candlepower is but slightly less than that of a 1000-watt 115-volt 12-inch searchlight (approximately 250,000 candlepower). Higher candlepowers, it is believed, would be but little more effective in penetrating smoke dense enough to permit only a few feet of visibility with a 200,000 candlepower beam since vailing brightness of the illuminated smoke particles tends to obscure vision as beam intensities are increased. If the smoke is dense enough or the distance great enough, overlapping particles present a physical obstruction between the eye and the object it is desired to see, and under such conditions, no amount of illumination will improve visibility.

4-103 The possibility of achieving greater light penetration by the use of selected wavelengths of light does not appear promising. There are but negligible differences in the transmissivity of water vapor to various wavelengths of monochromatic light and it seems doubtful that greater differences would be found to exist in the case of smoke. Subtractive filters must be used with tungsten filament pumps to obtain monochromatic light and these filters greatly reduce the intensity of the light produced by the filament. It is not considered likely that any wavelength of monochromatic light could offer enough advantage over white light so far as smoke penetration is concerned to offset the loss of intensity that would be caused by filter absorption.

(4) Battle Dressing Stations

4-104 Battle dressing stations as presently located on the hangar and gallery spaces on CV9 Class carriers are extremely vulnerable to damage and subject personnel stationed therein to unnecessary hazard. In cooperation with the Bureau of Medicine and Surgery, the problem was reviewed and ShipAlt CV809 of 27 June 1945 authorized the following changes:

1. Relocation of auxiliary battle dressing station No. 4 from hangar deck at frame 85, starboard, to centerline in passage A-105-L, hangar deck, frames 12-15.
2. Relocation of auxiliary battle dressing station No. 5 from hangar deck at frames 155-158, port, to hangar deck at fantail abaft bulkhead 202-1/2.

3. Relocation of emergency battle dressing station in B-0207-L, frames 107-111, gallery deck, to port of centerline at frames 74-77, on gallery deck.

(5) Aircraft Jettisoning Stations

4-105 The desirability of providing several widely separated aircraft jettisoning stations on the hangar deck has been demonstrated by war experience. As designed and built, CV9 Class carriers were able to jettison aircraft only at the deck-edge elevator opening, port side, frames 93-98. Hangar fires have generally been of such proportions that this amidships station has been engulfed in flames, rendering it impossible to jettison any aircraft even though some might be parked outside the fire area and therefore accessible for

--49--

jettisoning. For such cases where aircraft parked forward and aft are approachable during the fire and jettisoning of some or all of these as a precautionary measure is desired, and also for disposal of damaged aircraft after fires have been extinguished, two additional jettisoning stations have been authorized at frames 42-57, starboard, and frames 165-177, starboard. The latter station requires the relocation of the existing boat stowage to a new position aft of the boat crane. Cutting of the sheer strake coaming flush with the hangar deck at both stations was authorized by reference (u). On CVE's and CVL/s the hangar deck openings are of insufficient size to permit provision of jettisoning facilities.

Section V - Conclusion

5-1 It is apparent from a study of the damage experiences of *Franklin* and other carriers that fire remains the major hazard to the safety and combat efficiency of aircraft carriers when subjected to suicide crashes and bombing attacks. Damage control efforts and training should be mainly directed toward the elimination of fire hazards and the rapid extinguishing of fire once it breaks out. All shipboard personnel including the air group should attend firefighters' schools and should be constantly drilled in the location and proper use of shipboard firefighting equipment. It is expected that the new high capacity fog foam systems currently being installed on carrier flight and hangar decks will provide the ship's force with a truly effective means for the control and rapid extinguishment of gasoline fires.

5-2 It now appears doubtful that the Japanese Kamikaze type of attack will again become a major factor in naval warfare but is rather a forerunner of the development and application of pilotless aircraft and guided missiles. Viewed in this respect, it is apparent that much additional information of possible future significance can be obtained by a careful study of the wealth of war damage experience resulting from Kamikaze attacks.

--50--

Caption: Photo 1: 30 October Action. Starboard side showing fire in way of plane crash.

Caption: Photo 2: 30 October Action. Flight deck looking starboard and aft, showing damage at frames 125-128.

--51--

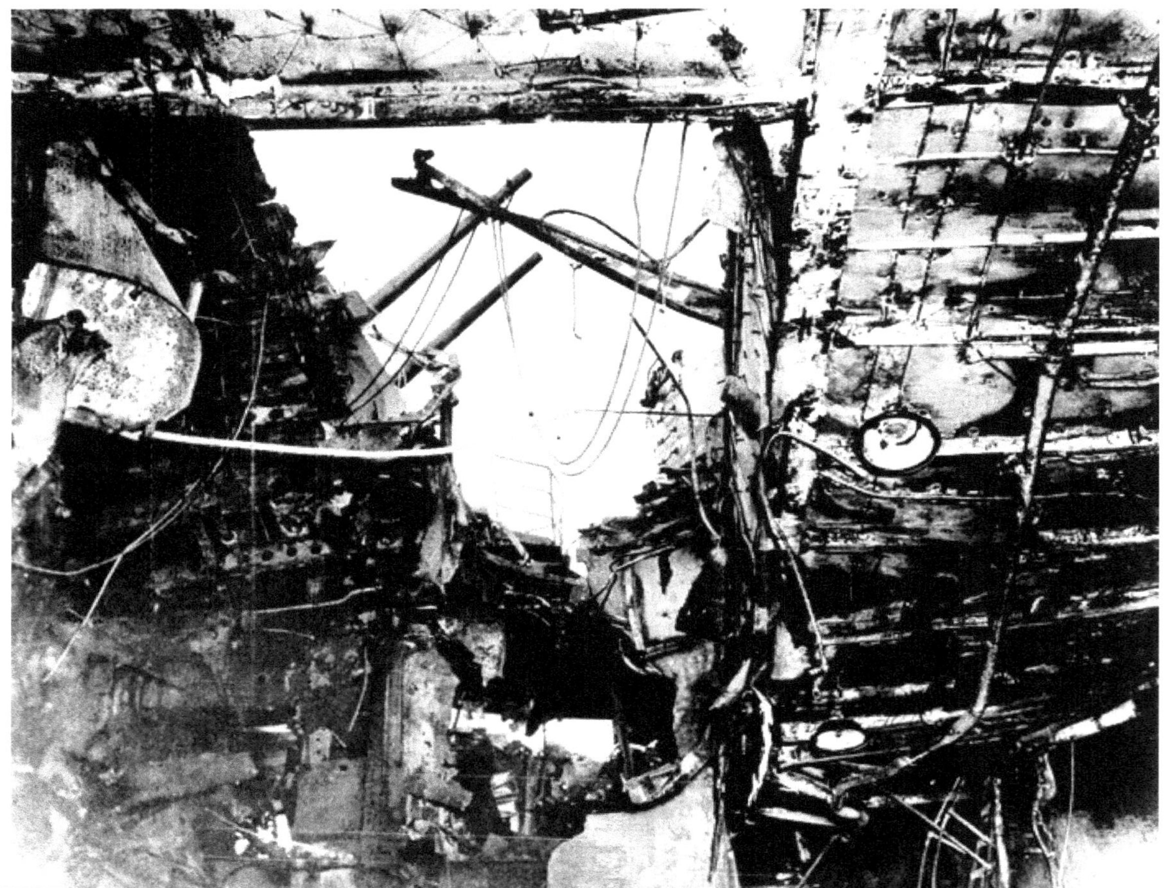

Caption: Photo 3: 30 October Action. View taken from hangar looking to starboard and up. Note damage to bomb elevator trunk and sprinkler equipment.

Caption: Photo 4: 30 October Action. View taken from flight deck looking to port and down into hangar.

Caption: Photo 5: 30 October Action. View taken from hangar looking to starboard. Note damage to air intake ducts; bomb elevator trunk; ladder.

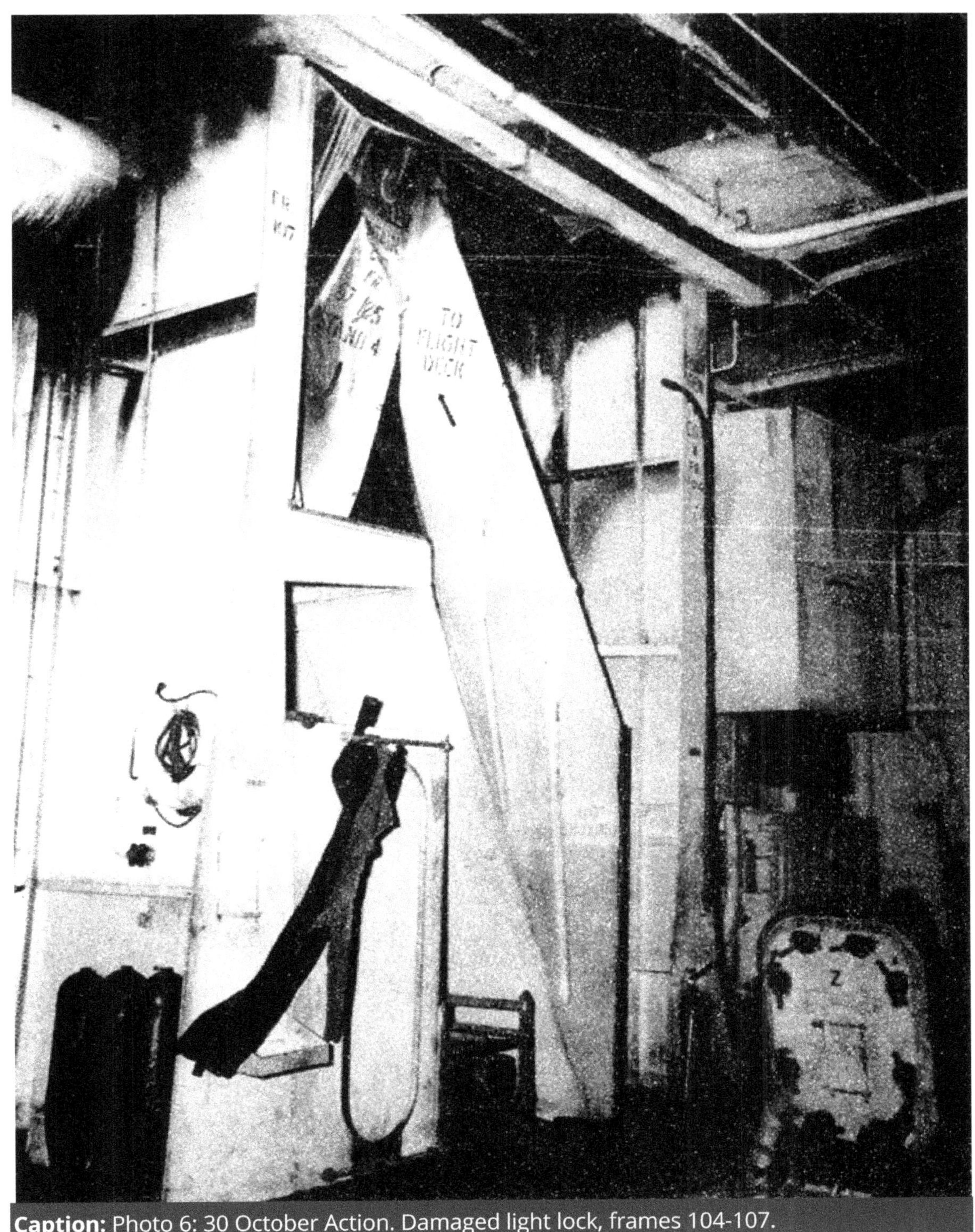

Caption: Photo 6: 30 October Action. Damaged light lock, frames 104-107.

Caption: Photo 7: 30 October Action. View showing distortion of uptake enclosure bulkhead on which hydraulic valve controls are mounted, frame 112, compartment B-313-E. Hydraulic controls remained operable.

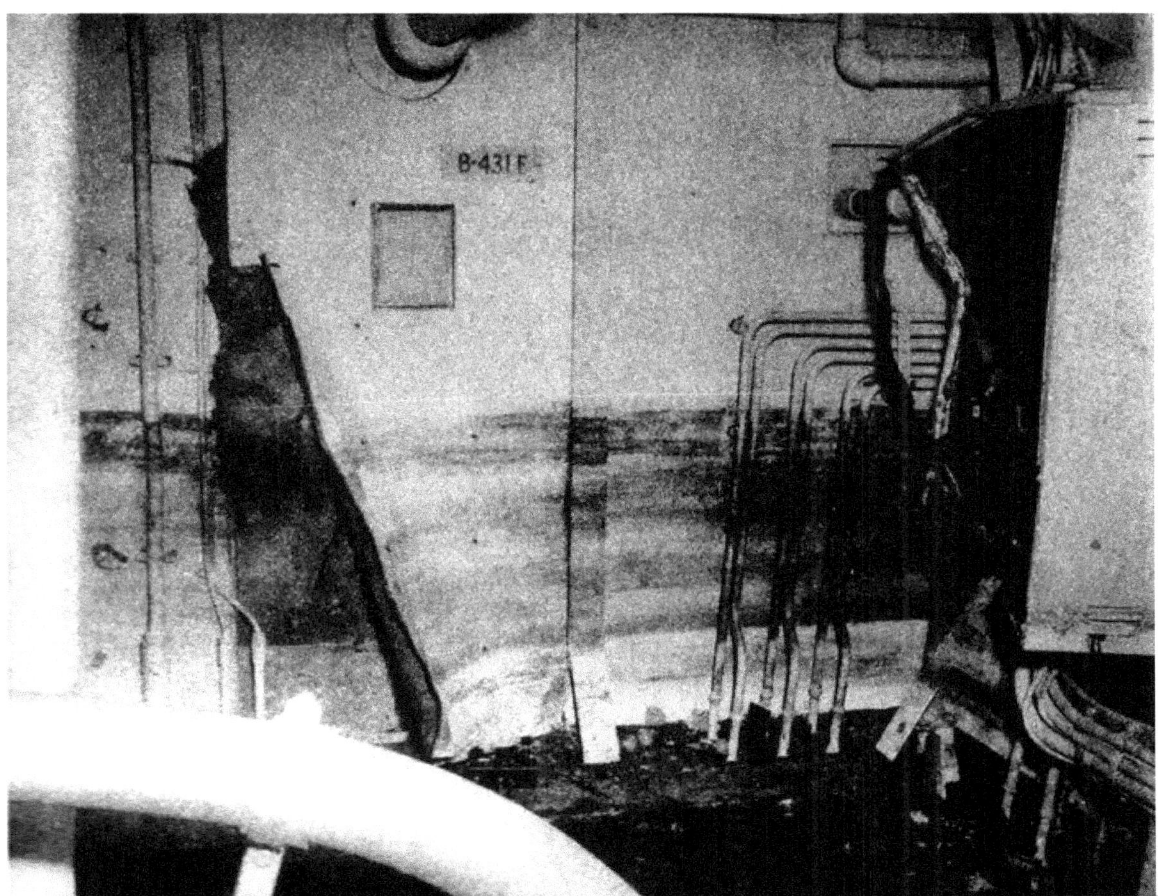

Caption: Photo 8: 30 October Action. View showing damage by vapor explosion to sheathing and bomb elevator machinery controller box in B-431-E. Note high-water mark.

Caption: Photo 9: 30 October Action. View showing damaged bomb elevator, B-0435-T, and damaged watertight door 3-128-1 as seen from B-318-L, crew's galley, looking forward.

Caption: Photo 10: 30 October Action. View showing dished-in after bulkhead of uptake enclosure B-315-E at frame 114-1/2.

Caption: Photo 11. 19 March Action. Distant shot showing black smoke of gasoline fire and white smoke of high order detonation.

Caption: Photo 12. 19 March Action. Franklin enveloped in smoke.

Caption: Photo 13: 19 March Action. A tremendous explosion. Note extent of high order detonation white smoke. Rockets and other pyrotechnics are distinctly visible at stern.

Caption: Photo 14: 19 March Action. Picture taken at the instant of a high order detonation on port side. Note debris in air and firefighters running to escape. Note topmast broken at radar platform level.

Caption: Photo 15: 19 March Action. Close-up of fire from starboard side. Intense fire seen aft of 40mm quad No. 15 is from gasoline draining from hangar via roller door opening, frame 170.

Caption: Photo 16: 19 March Action. Distant shot of fire in after part of Franklin. Burning gasoline is still pouring out of roller door opening about frame 170. Note men at stern escaping via net.

Caption: Photo 17: 19 March Action. Fire is well aft but some burning gasoline is still draining from starboard side of hangar.

Caption: Photo 18: 19 March Action. USS Santa Fe (CL60) alongside.

Caption: Photo 19: 19 March Action. Ammunition in No. 7 twin mount is burning. This occurred late in fire. Note that fire on flight deck aft is about out.

Caption: Photo 20: 19 March Action. View of island looking forward and to starboard.

--60--

Caption: Photo 21: 19 March Action. View looking forward at after expansion joint. Note damage to after elevator platform.

Caption: Photo 22: 19 March Action. Large hole port side, frames 118-130, caused by bomb which dropped down to about gallery deck level.

Caption: Photo 23: 19 March Action. View of hole in edge of flight and gallery decks on port quarter, frames 201-207.

Caption: Photo 24: 19 March Action. Holes cut in flight deck forward to extinguish fires in gallery deck spaces.

Caption: Photo 25: 19 March Action. General view of flight deck aft as Franklin arrived at Navy Yard, New York.

Caption: Photo 26: 19 March Action. After end of flight deck.

Caption: Photo 27: 19 March Action. General view of hangar looking aft from about frame 112.

Caption: Photo 28: 19 March Action. After elevator.

Caption: Photo 29: 19 March Action. Forward elevator.

--64--

Caption: Photo 30: 19 March Action. Looking to port and up from hangar to the platform for Nos. 6 and 8 5-inch guns on port quarter. Large hole blown in gallery deck space by bomb.

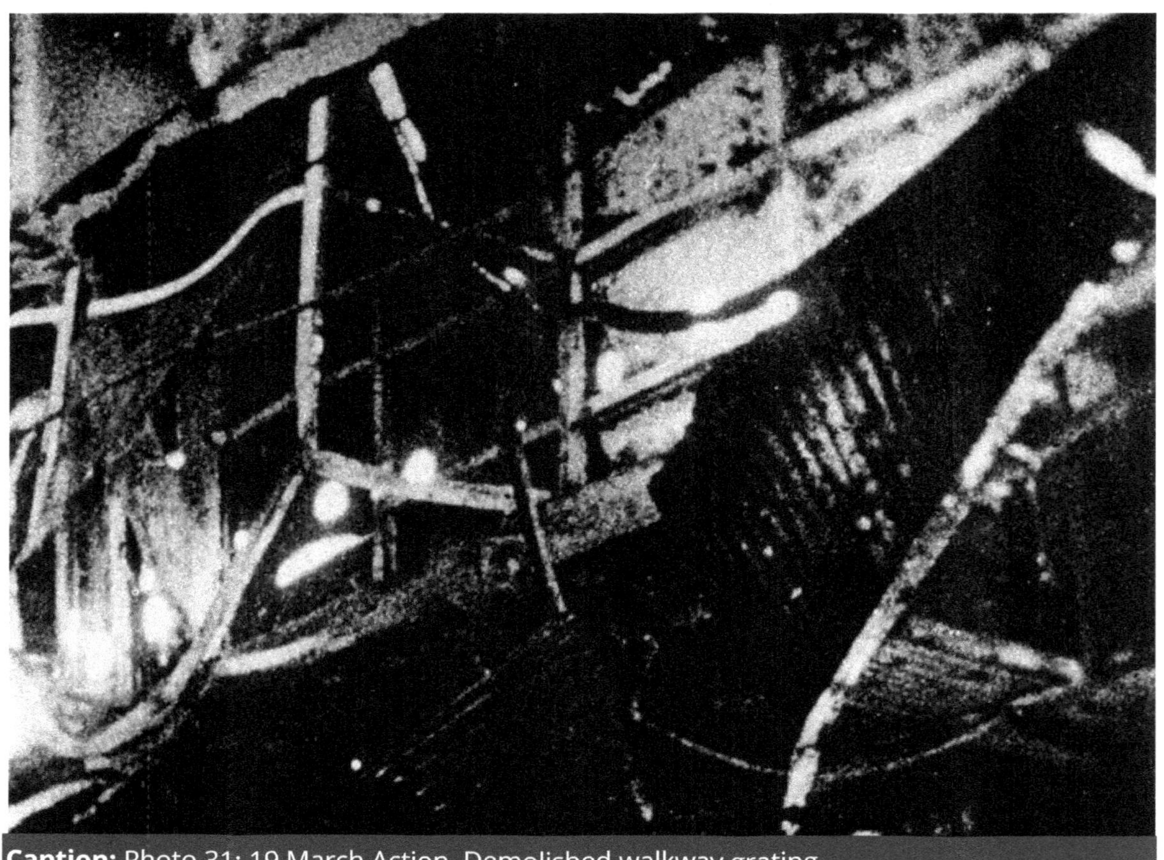

Caption: Photo 31: 19 March Action. Demolished walkway grating.

Caption: Photo 32: 19 March Action. Hangar looking to starboard, frame 79.

Caption: Photo 33: 19 March Action. Hangar view showing moving picture booth and bomb elevator trunk.

Caption: Photo 34: 19 March Action. View of demolished ladder in hangar.

Caption: Photo 35: 19 March Action. Hole in hangar deck, frame 100, caused by Tiny Tim rocket. Note fragmentation pits on STS uptake bulkhead.

Caption: Photo 36: 19 March Action. Bomb case. Low order detonation.

Caption: Photo 37: 19 March Action. Tiny Tim rocket casing lodged in after bulkhead (1¼-inch STS) of after elevator pit.

--68--

Caption: Photo 38: 19 March Action. CIC, on gallery deck, showing extensive damage by blast and fire.

Caption: Photo 39: 19 March Action. Upper handling room No. 7 5-inch twin mount, starboard side looking forward. Forward starboard hoist is intact.

Caption: Photo 40. 19 March Action. Upper handling room No. 7 5-inch twin mount. Looking outboard to starboard. Hoists and center column destroyed by explosions and fire. Note hole in 3/4-inch STS outboard bulkhead.

Caption: Photo 41. 19 March Action. Upper handling room No. 7 5-inch twin mount. Looking aft port side. Note damage to after port hoist.

Caption: Photo 42: 19 March Action. Upper handling room No. 7 5-inch twin mount. Looking to port.

Caption: Photo 43: 19 March Action. Upper handling room No. 7 5-inch twin mount. Looking to starboard. 3/4-inch STS deck and after bulkhead split open.

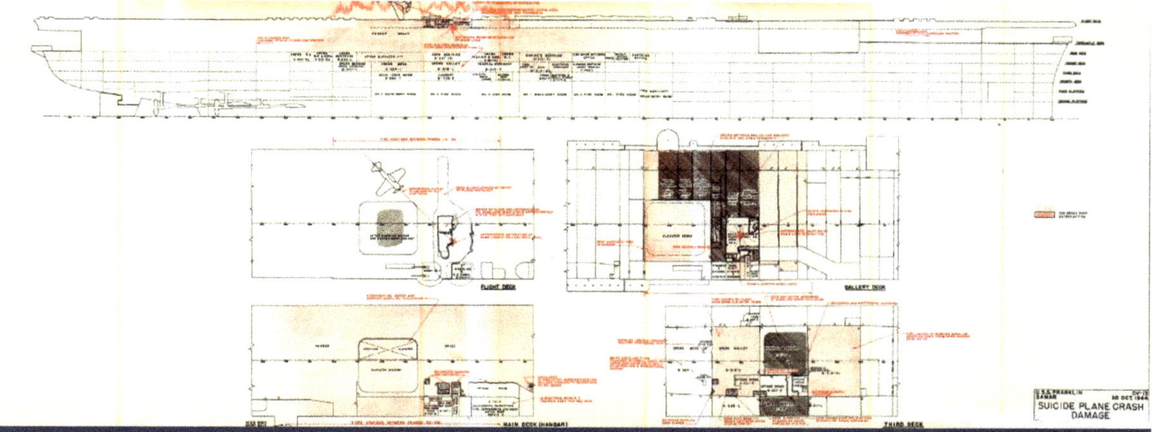

Caption: Plate 1: Suicide Plane Crash Damage.

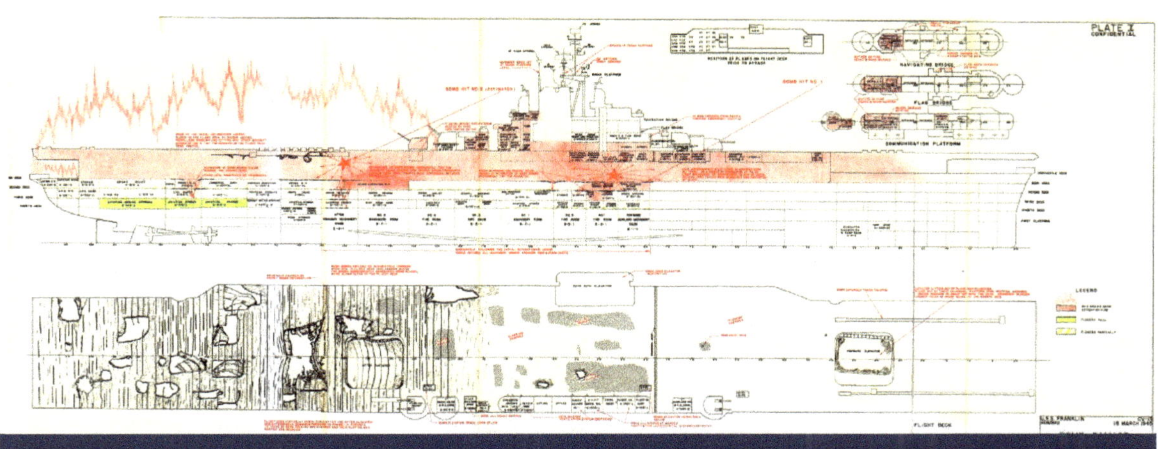

Caption: Plate 2: Bomb Damage.

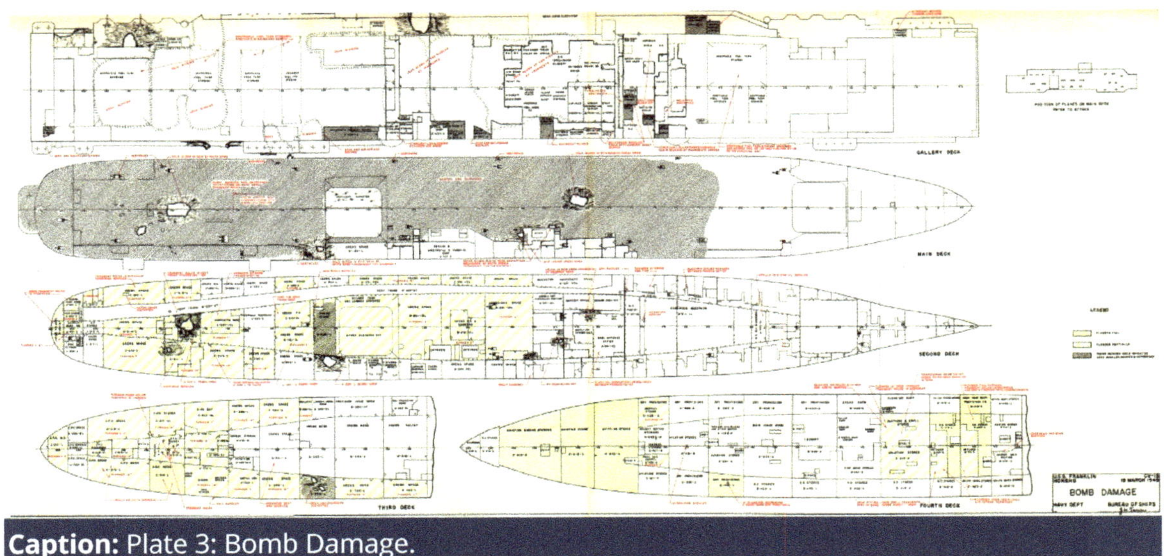

Caption: Plate 3: Bomb Damage.

U.S.S. WASP (CV7)

Loss in Action

South Pacific
15 September,1942

Class............Aircraft Carrier	Length (W.L.)......690'-0"
Launched.............April 4,1939	Beam (W.L.)........ 80'-8"
Displacement..........14,700 Tons (Standard)	Draft.............. 22'-2" (Designed)

Reference:

(a) C.O. WASP ltr. P6-1 (006) of 24 September, 1942 (Loss Report).

CONTENTS

LIST OF PHOTOGRAPHS

LIST OF PLATES

SECTION I - SUMMARY

1. WASP (CV7) was torpedoed by a submarine during the afternoon of 15 September, 1942 while operating with a task force southeast of the Solomon Islands. When attacked, planes were being rearmed and refueled, and a large portion of her gasoline system consequently was in use. Two torpedoes struck and detonated against the starboard side, one about frame 36 in way of the gasoline stowage tanks and the other about frame 62 in way of the forward bomb magazines. About 20 seconds after the torpedoes detonated a third explosion, of somewhat less intensity, occurred. Immediately following the torpedo detonations and the explosion, gasoline fires broke out in the vicinity of the athwartship gasoline main on the second deck at frame 63 and low in the vessel in the vicinity of the ruptured gasoline tanks. In addition to these fires another serious conflagration was started in the forward portion of the hangar. Free gasoline pouring from the ruptured tanks onto the surface of the water was ignited, and soon WASP was completely enveloped in flames from frame 20 to frame 80. Within 25 minutes of the attack, three additional major gasoline vapor explosions occurred. Very little water pressure was available at the fire plugs on the hangar and flight decks. The order to abandon WASP was given about 35 minutes after the torpedoes struck. The vessel burned throughout the rest of the day and was sunk by U.S. destroyer torpedoes at nightfall.

2. With the exception of the forward diesel generators and the No. 1 main generator in the forward engine room, the engineering plant was intact and engineering personnel remained at their stations until WASP was abandoned. These circumstances, plus the fact that it was necessary for U.S. forces to complete WASP's destruction, are clear indications that structural damage and loss of buoyancy and stability were not fatal. The loss of the WASP thus must be attributed to the fires which resulted from the ignition of gasoline.

3. WASP was a new vessel, completed in 1940, and is considered to have been provided with an excellently arranged fire main system with ample pumping capacity. The inability to control the fire on the hangar deck is disappointing in view of the excellent facilities provided and the intact condition of the pumping plant. It was this fire which eventually forced WASP to be abandoned.

4. It appears that initially there were three distinct areas in which fires occurred. These areas were first, the hangar; second, the second deck in the vicinity of frame 63; and third, low in the vessel in the vicinity of the ruptured gasoline tanks. The fire on the second deck resulted from gasoline draining into this area from the mains above and on the outside of the ship. The fire low in the vessel came from free gasoline from the ruptured tanks. The fire in the hangar was undoubtedly caused by aircraft falling from the overhead onto other aircraft parked on the deck rupturing aircraft gasoline tanks and allowing gasoline to escape. After consideration of these factors, it has been concluded that the fire in the hangar, together with low fire main pressure on the plugs in the hangar, was primarily responsible for the

loss of WASP. If the hangar had been available as an area for damage control operations and had remaining facilities been utilized, it seems probable that the fire on the second deck would have been controlled quite readily, inasmuch as the quantity of gasoline feeding this fire was not very great. Whether the fire low in the ship could have been confined and eventually extinguished, had the opportunity been available to exert efforts to this end, is problematical. Other cases of fighting severe oil and gasoline fires indicate a probability that this might have been accomplished. On the other hand, severe gasoline vapor explosions might have prevented confinement of the fire.

5. The references report very low pressure at the fire plugs on the hangar deck. Some survivors have attributed this lack of pressure to destruction of the fire main by shock. After a thorough study of all the evidence, the conclusion was reached that the low pressure on the hangar deck fire plugs was caused by an open break in the forward loop of the fire main, and that the forward loop was never isolated from the balance of the system.

6. Thus the circumstances involved in the loss of WASP were somewhat similar to those which caused the loss of LEXINGTON (CV2)* inasmuch as both vessels had to be abandoned by reason of the raging internal uncontrolled fires followed by internal explosions. In neither case were the structural damage and loss of buoyancy and stability caused by the torpedoes sufficient to have caused the loss of the vessels.

7. The case of LEXINGTON (CV2) indicated a need for further improvement in gasoline stowage and handling arrangements for carriers, and the case of WASP emphasized this need. Shortly after the loss of LEXINGTON a program of research was inaugurated and, at the same time, an investigation was made of merchant marine and industrial practices for the stowage and handling of gasoline. The improvements and changes which are being placed into effect are discussed in some detail in Section III. It is hoped that a marked increase in resistance to damage of the gasoline system and ease of handling gasoline will result. Most of the improvements discussed in Section III will be incorporated in the later vessels of the ESSEX (CV9) class and the CVB41, new CVL and CVE105 classes of carriers.

SECTION II - NARRATIVE
(Plates I and II, All Photos)

8. The data on which this report is based were obtained from the Commanding Officer's report, reference (a). That reference included statements by several of the surviving officers, and is a more complete description than is usual in such cases of severe damage followed by loss. The extent of structural damage, however, is not of record and apparently was largely undetermined. The Bureau accordingly has supplemented the data contained in reference (a) with an estimate of the probable extent of structural damage. This estimate is based on a study of a number of cases of torpedo damage to vessels with underwater layouts similar to that of WASP. Plates I and II thus were prepared from a combination of the known facts and estimate based on other war experience. The

---

* BuShips War Damage Report No. 16

photographs are copies of pictures taken by the various photographic units attached to the task force. Plate III is a diagrammatic arrangement of the new type of gasoline stowage discussed in Section III.

9. During the afternoon of 15 September, 1942 WASP was a unit of one of two carrier task forces operating jointly in an area southeast of the Solomon Islands. The sea was slight with a few whitecaps. The wind was about 14 knots from east southeast. The formation's base course was 280 degrees. WASP was duty carrier, periodically turning into the wind to launch and recover aircraft. The ship was in Condition II with the air department at flight quarters. Material condition "BAKER ALERT" was set. At 1420 WASP turned into the wind on course 120 degrees and launched 18 VSBs and 8 VFs. Immediately after the launching was completed the ship took aboard 8 VFs and 3 VSBs. Upon completion of recovery, the ship commenced to turn right with standard rudder at 16 knots to base course. This was at 1442. During flight operations refueling and rearming of aircraft on board were in progress. Torpedo planes in the hangar were armed with torpedoes but gasoline tanks had not been refilled and still contained CO_2. Scout bombers in the hangar were fully fueled and armed with depth bombs. Fighters on both the flight deck and in the hangar were fueled and armed. The gasoline system was in use. Orders had been given to re-spot aircraft and, in connection with this, the forward section of the gasoline system was being secured and the hoses were being walked back to drain gasoline into the tanks. This, however, had not been completed at 1445. At the same time the after section of the gasoline system was being prepared for use and gasoline was being pumped through the mains. It is not known whether fuel had reached the flight deck at 1445.

10. At 1445, while in a turn to the right, two submarine torpedoes were sighted close aboard approaching from about three points forward of the starboard beam. They appeared to have been fired from a point close to the vessel as they were still rising when sighted. They were estimated to have been between 10 and 15 feet below the surface when they struck and detonated. The first one hit at about frame 36, starboard. The second struck almost immediately afterward at about frame 62 just forward of the island structure. The ship immediately listed 15 degrees to starboard and finally came to rest at 11 degrees. Flooding resulted in a large trim by the bow. An examination of the photographs indicates that this trim was about 10 or 12 feet.

11. The first torpedo which struck the starboard side at approximately frame 36, destroyed the outer and inner skins of the vessel and ruptured gasoline tanks in A-10-Gas. The 5-inch handling room A-602-M was undoubtedly flooded as was probably the 5-inch powder magazine A-601-M immediately forward. The first platform was probably disrupted over a large area and the storerooms on this deck were flooded. The fourth deck was also probably ruptured as survivors abandoning the forecastle noted a hole in the starboard shell above the waterline despite the fact that the ship was down by the bow and was listing to starboard. The gasoline suction and discharge mains running under the first platform were probably destroyed

or broken as far aft as frame 45. The gasoline trunk A-512-T and the gasoline pump motor room were probably wrecked. This torpedo also may have ruptured the boundaries of the gasoline tanks in A-13-Gas. Gasoline quickly flowed from A-10-Gas onto the surface of the water above the hit. Gasoline was also probably sprayed through damaged first platform compartments.

12. The second torpedo, which struck almost immediately after the first, destroyed the inner and outer skins on the starboard side and flooded the bomb magazines in the area between frames 56 and 68. Bulkhead 62 was probably destroyed on the starboard side at least to the fourth deck. The first platform was undoubtedly disrupted, permitting the flooding of the storage spaces on this level from side to side between frames 49-1/2 and 76. The centerline bulkhead between frames 56 and 68 on the second platform was probably ruptured, flooding the bomb magazines on the port side. Bulkhead 68, separating the bomb magazines from the plotting room and central station was ruptured. The plotting room flooded rapidly with water and oil, and central flooded slowly through trunk A-619-T. Undoubtedly, the gasoline tanks in compartment A-13-Gas forward of this hit were also ruptured, and the after gasoline trunk A-521-T was damaged. The 1".1, 20mm, .50 and .30 caliber ammunition magazines in the hold below the second platform were probably flooded in way of this hit. The overall extent of damage and flooding was not reported but was probably about as shown on plates I and II.

13. Undoubtedly, the forward loop of the fire main, as shown on plates I and II, was ruptured at least in the starboard section in way of the two torpedo detonations. This made inoperative the fire main risers connected to the forward loop.

14. Survivor reports indicate that shock from the two torpedo hits was considerable. A ladder on the main deck about frame 20 was torn from its fastenings and hung straight down from the forecastle deck hatch. The four aircraft triced to the overhead forward fell and struck the planes parked on the hangar deck. The landing gears on all aircraft collapsed. This damage was evidently due to shock as no serious blast effects occurred in the hangar at this time. The forward diesel generators were thrown from their foundations and all lights went out in the forward half of the vessel. In the forward engine room the forward main distribution board was damaged. In addition, the shell of the forward dynamo condenser was reported to have split, permitting steam to escape into the forward engine room. Shock damage was also reported to instruments located in the control spaces in the island. Flexural vibration must also have been present inasmuch as whipping of the foremast was reported, which dislodged numerous rivets.

15. Simultaneously with the impact of the second torpedo hit, flames appeared on the water in way of the forward hit at frame 36. A short time later flames also appeared in the forward section of the hangar. This fire in the hangar was probably started by gasoline spilling from the tanks of the aircraft. Rupture of the tanks probably was caused by falling planes and debris. This fire quickly spread to the starboard

5-inch gun platform. A short time later ready service projectiles for 5-inch gun No. 3 began to detonate.

16. About 20 seconds after the torpedoes struck, a third explosion - somewhat less severe - was felt. This was reported by the Commanding Officer to have been a torpedo, but it is believed to have been of internal origin inasmuch as the flight deck cover to bomb elevator A-423-ET was blown violently upward and flames appeared in the shaft. The cause of this explosion is unknown, but possibly gasoline vapors from the gasoline tank A-13-Gas passed through the bomb magazine and into the bomb elevator trunk A-711-ET where they might have collected. Or possibly, they might have collected in the bomb arming stations A-309-L and A-310-L. The explosion was witnessed by a survivor who was on the first platform deck in trunk A-619-T. He saw the flash of the explosion on the third deck and was thrown down the trunk to the next deck. The blast apparently traveled up the bomb elevator trunk A-423-ET and blew off the bomb elevator hatch cover on the flight deck. Other blast effects were felt throughout the island structure and spaces on the second deck. Paint on the after side of bulkhead 76 in the log room, which is immediately aft of the bomb arming station, caught fire. The athwartships gasoline mains under the main deck at frame 63 were reported to have leaked, probably at the fittings, and the spilled gasoline burned. Whether the leaks in gasoline piping were caused by the vapor explosion or the torpedo detonation is open to question, but the Bureau considers that the vapor explosion is a more probable cause of this leakage. The second deck may have been ruptured by this explosion. The aluminum bulkheads on the starboard side were collapsed and disrupted as far aft as frame 87. The area between frames 62 and 76 on the second deck was soon on fire completely across the vessel. The after section of the gasoline system, which at the time of the torpedo hits was being placed into use, contained gasoline. These mains probably drained by gravity into the transverse gasoline mains at frame 63 on the second deck, inasmuch as the references contain no mention of attempts to cut out this section by the cut-out valves provided in the mains. Since this main was leaking, a constant source of gasoline was furnished for some moments to the vicinity of frame 63 on the second and the third decks. Three serious vapor explosions occurred in this area prior to the ship's abandonment.

17. The first of these explosions, "A" on Plates I and II, occurred at a time variously estimated to be from three to five minutes after the torpedoes hit. It evidently centered in the area around the athwartship gasoline main at frame 63, probably in the vicinity of the area disrupted by the previous explosion in the bomb arming station below. Structural damage from this explosion was not reported, but the shell was not torn open. The blast traveled up the already damaged bomb elevator trunk and sent flames and heavy smoke up the inboard side of the island structure. The bulkheads separating the No. 2 elevator pit from the scene of the explosion were probably disrupted as the blast lifted the elevator, which was at the flight deck, sending the light waterway cover plates about 20 feet into the air.

18. At 1505, 20 minutes after the torpedoes struck, the second heavy vapor explosion "B" occurred. It probably

centered in the third deck area in or adjacent to the explosion in the bomb arming station. This explosion manifested itself above the flight deck in a large ball of flaming gases which completely enveloped the forward portion of the island. These gases apparently came from the third deck area through the previously damaged bomb elevator trunk. The No. 2 1".1 quadruple gun mount immediately forward of the bridge was blown from its base. Twenty millimeter and 1".1 cartridges in the clipping rooms in this area started to go off with low order detonations because of the heat from the surrounding fires. The bridge was abandoned at this time.

19. The third major vapor explosion "C" occurred at 1510. Its effects were particularly noted in the hangar deck area. No. II elevator was lifted bodily and fell back on the flight deck so that it lay askew across the elevator opening. Steel plating, expansion joint covers and barrier stanchions around elevator No. II were blown as high as 150 feet into the air. An officer who was supervising the firefighting activities on the hangar deck was blown from his position aft of No. II elevator about 50 feet along the deck. The dense smoke from this explosion was noticed by the screening vessels. The center of this explosion was probably in the elevator pit on the second deck. This area had been wrecked by previous explosions.

20. The fire had already spread to the starboard forward 5-inch gun group and caused the detonation of some ready service projectiles for these guns. The bulkhead separating this platform from the interior of the ship was riddled and the crew's space A-0206-L was wrecked. Bulkheads and spaces forward and aft of this platform were ripped and torn. Bulkhead 25 separating the hangar from the wardroom country was reported to have been destroyed. The fire rapidly crossed through the ship and eventually resulted in the detonation of the ready service projectiles in the forward port 5-inch gun group. The whole area below the flight deck between the forward port and starboard gun groups was so completely wrecked that a large hole through the vessel was noted by the Air Officer after he abandoned the ship.

21. There were 16 aircraft, four SBD-3's and 12 F4F-4's on the hangar deck forward of No. 2 elevator. All were fully loaded with gasoline and .30 and .50 caliber cartridges. In addition, the four SBD-3's were armed with one 325-pound depth bomb each. The fires in the hangar forward of No. II elevator resulted in the .30 and .50 caliber cartridges exploding almost continuously until the ship was abandoned. The gasoline and depth bombs undoubtedly contributed immensely to the intensity of the fire. The depth bombs apparently did not detonate as no violent explosions were reported in the forward section of the hangar.

22. It is noteworthy that no evidence was reported in the reference which would indicate that a mass detonation of the bomb magazines or of the ready-service stowages for 5-inch and 1".1 guns occurred.

23. The fire main system on WASP was arranged in three loops with one loop forward of the main machinery spaces, one

loop in the main machinery spaces and one loop aft of the main machinery spaces. Cut-outs were provided in the engine rooms to permit isolation of any one loop from the others. Since the torpedoes detonated on the starboard side forward, the starboard section of the forward loop undoubtedly was ruptured. Thus, all risers from this loop were ineffective. The risers from the center loop were probably undamaged as evidenced by the fact that weak streams were reported at fire plugs at frame 98 and frame 105 on the hangar deck. Attempts were also made to use the fire plugs at frame 100 and frame 106 on the flight deck, but no pressure was available. The low pressure at the plugs was undoubtedly due to insufficient pressure in the system rather than damage to the risers in use inasmuch as there is no evidence that structural damage occurred aft of frame 100 until later.

24. Throughout the whole action power was not lost at any time, although difficulty was experienced in maintaining the feed water supply to the after boilers. Pressure was lost rapidly from the after part of the feed system through a leak which was never located. Large quantities of make-up feed water had to be pumped up from the bottoms and both of the main feed pumps aft had to be run at maximum capacity in order to maintain operating pressure. Emergency feed pumps in the boiler spaces were kept in a standby condition.

25. During the period discussed above, steering control was aft, orders being transmitted from the bridge via the JV circuit. Immediately after WASP was hit, the rudder was changed to full left with the object of utilizing the wind to blow the fire away from the undamaged portion of the ship, and also to clear the burning gasoline and oil on the surface of the water. This was successful inasmuch as the wind was maintained on the starboard quarter. This prevented the surface fire from spreading aft, however, attempts to back clear of the burning gasoline and oil were unsuccessful because the gasoline and oil continued to rise to the surface from the ruptured tanks.

26. WASP settled with a list of about 11 degrees to starboard immediately after the torpedoes struck. Steps to reduce this list were taken immediately. Fuel oil was transferred from fuel oil tanks B-23-F and C-711-F to B-24-F either by the fuel oil transfer and booster pumps or the after damage control pump in the after engine room. In addition, tanks C-708-F and C-710-F were filled from the sea through the damage control main. This admitted about 153 tons of water. By the time the ship was abandoned, 35 minutes after the torpedo hits, the list was reduced to about four degrees to starboard.

27. By 1515 the fires had spread throughout the forward half of the ship and efforts to control them were futile. The Commanding Officer ordered WASP abandoned at 1520 to prevent unnecessary loss of life. This was completed at 1600. After the ship was abandoned, the fires traveled aft rapidly and several violent explosions occurred. These were reported by LANSDOWNE and SALT LAKE CITY to have occurred at 1627, 1746, 1757 and 1758. Throughout this period WASP's list gradually increased to about 15 degrees and the hangar

deck forward was awash. At nightfall LANSDOWNE was ordered to sink WASP. Between 1907 and 2010 five torpedoes were fired. Two hits were obtained on the starboard side, both amidships, and one on the port side just abaft the island structure. The fire by now had completely enveloped the stern. Both the ship and the pool of gasoline and oil in which she was floating were burning with great flames, illuminating a large area. At 2100, while listing heavily to starboard, WASP sank bodily by the bow.

28. Summarizing, it appears that the following events took place aboard WASP at the times stated:

1445 - Two torpedoes struck the starboard side almost simultaneously about frames 36 and 62.

1445 - Internal explosion (20 seconds later) - possibly in the vicinity of the bomb arming station.

1449 - First gasoline vapor explosion "A" - under inboard edge of island, probably on second deck.

1505 - Second gasoline vapor explosion "B" - under forward section of island on either second or third deck.

1510 - Third vapor explosion "C" - in vicinity of No. 2 elevator pit.

1450-1520 - Numerous minor explosions - ready service projectiles, aircraft gasoline tanks and machine gun cartridges.

1520 - Hangar in flames forward of No. 2 elevator. Ship ordered abandoned.

1600 - Abandoning completed - fire working aft - minor explosions continuing.

1627 - Large explosion - noted by LANSDOWNE.

1746 - Extremely violent explosion - noted by LANSDOWNE and SALT LAKE CITY.

1757 - Large explosion - noted by SALT LAKE CITY.

1758 - Large explosion - noted by SALT LAKE CITY.

1908-2011 - Five torpedoes fired by LANSDOWNE - three hits.

2011 - Fires spread to stern.

2100 - WASP sank bodily, going down by the bow.

SECTION III - DISCUSSION

A. Type of Torpedoes

29. Two types of 21-inch Japanese submarine torpedoes are known to exist. The larger contains 660 pounds of hexa in

the warhead and the smaller is estimated to carry a charge of 450 pounds of hexa. In view of the fact that WASP was in no immediate danger of sinking as a result of flooding, structural damage could not have been unusually extensive. The warhead charges employed could not well have been larger than 660 pounds and possibly were as small as 450 pounds.

B. Damage Control Measures

30. The starboard list of 11 degrees was quickly reduced to 4 degrees by transferring fuel oil and counterflooding two empty fuel tanks. Some 20 minutes after the attack these measures were completed.

31. It is noted however, that two fuel tanks, C-708-F and C-710-F, which were required to be kept full of either fuel oil or salt water at all times inasmuch as they constitute a portion of the liquid protection for the after magazines, were empty. The reason for their being empty was not reported. Although in this instance their empty condition provided rapid counterflooding capacity, it is nonetheless extremely dangerous to leave gaps in the underbody liquid layer. A liquid layer surrounding the magazines is provided in part for the purpose of reducing the effects of fragments following underwater explosions. Almost all high explosives are subject to detonation by fragments, and the danger of a magazine explosion is therefore greatly reduced when a liquid layer is provided. In this connection the magazine explosion which caused the loss of the bow of NEW ORLEANS* resulted from a torpedo detonation against the shell plating of a bomb magazine where no liquid layer was provided. Very probably the mass detonation of the bombs was caused by fragments of the torpedo and ship structure produced by the torpedo explosion.

32. As noted in the summary, the inability to control the fire in the hangar deck was one of the vital factors which caused the entire situation to become hopeless in such a short period of time. The Commanding Officer reported that lack of pressure at the hangar deck fire plugs prevented effective firefighting measures. The question of pressure on the fire main is discussed in considerable detail in part E of this section. It is noted, however, that the reference does not contain any mention of attempts to cut out the damaged portion of the fire main forward of the engine room or the portion of the gasoline mains aft of frame 63. The fire on the second deck was undoubtedly fed by gasoline draining through local fractures in the main in the vicinity of frame 63. It is also probable that vapor from the main gasoline tanks contributed to fires and vapor explosions in this area. The gasoline pumps had been started just prior to the time of the torpedo attack in order to fuel planes on the after portion of the flight deck. Although these pumps were stopped by the torpedo detonation, possibly the mains and risers were full of gasoline at the time of damage. If the mains and risers were full, not more than 300 gallons of gasoline were available to feed the fires in the vicinity of the break at frame 63. This quantity of gasoline would drain through the leaks quite

- -

*BuShips War Damage Report No. 38.

slowly. This probably accounts for the appreciable time interval between the vapor explosions which occurred in the second and third deck areas.

33. It appears that some chance of successfully controlling the gasoline fire on the second deck and of confining the fire in the vicinity of the gasoline tanks would have existed had the fire on the hangar deck been controlled and extinguished within a comparatively short length of time. The fact that the hangar was aflame in the forward portion effectively prevented any efforts to control and isolate the fires lower down in the vessel. It is problematical, however, whether these lower gasoline fires, being fed from the main gasoline tanks and with frequently recurring vapor explosions, could have been extinguished even with full use of the facilities available to WASP. These facilities, as discussed later herein, included an excellent fire main system, hangar sprinkling and water curtain systems, but did not include later developments described in section E, including portable and fixed fog nozzles.

34. The forward repair party locker and headquarters were located on the third deck in compartment A-306-1L between frames 37 and 43. The damage deep in the ship killed the personnel at this station, and surrounding fires prevented access and use of its equipment. The question of proper location of repair parties with the necessary communication facilities and equipment is one in which several conflicting opinions need to be reconciled before it can be definitely settled. In the case of WASP damage control efforts in general and firefighting efforts in particular undoubtedly would have been better coordinated had the repair parties, or at least one or two of them, been located on the hangar deck. There has also been a suggestion that a secondary damage control station, with minimum essential facilities, be provided in cruisers and larger ships. This proposal is being given careful consideration. It appears that such a station might have been of considerable benefit in this case in providing a means for coordinating damage control efforts within the machinery spaces with those outside the machinery spaces. Specifically, it is quite possible that the fire main was not isolated because the personnel in the forward engine room had not received definite information as to the damage or definite instructions to isolate the fire main.

C. Machinery Damage

35. Damage to the main propulsive machinery did not occur. The main machinery remained in operation until orders to abandon ship were received. Damage from shock did occur to some of the auxiliaries. The forward diesel generators were thrown from their foundations when the two torpedoes detonated. The forward distribution board was also damaged. In addition, it is reported the forward dynamo condenser shell split and that steam escaped into the forward engine room. The shell of this condenser was made of 1/4-inch steel marine boiler plate with welded seams. The details of this failure were not reported. This is the first instance reported of failure

of a condenser shell due to shock. This failure, although extremely inconvenient to operating personnel in the forward engine room, did not prevent the operation of the main propulsion machinery, fire pumps and some of the other auxiliaries.

D. Fires and Explosions

36. The first interior explosion occurred about 20 seconds after the two torpedo hits. This explosion is reported by the Commanding Officer to have been caused by a third torpedo. There were four survivors however who, as eyewitnesses, reported seeing the wakes of only two torpedoes. Vessels in the screen also reported observing only two torpedo detonations. In addition, it is believed that flooding and structural damage would have been much more extensive than appears actually to have been the case, if a third torpedo had struck and detonated. An analysis of the reports of survivors who were in below-deck spaces indicates that this third explosion occurred on or below the third deck in the vicinity of the after torpedo hit, and was of considerably less intensity than the two torpedo detonations. This third explosion probably was a gasoline vapor explosion. Since the second torpedo had detonated about frame 62, the gasoline tanks in A-13-Gas were undoubtedly torn open. The gasoline vapors then would have been quickly disseminated throughout the damaged area and up the vertical trunks possibly into the bomb handling stations in A-309-L and A-310-L. Ignition of the vapor could have occurred from any one of a number of sources. The conjecture has been advanced that this third explosion was a bomb detonation in the bomb arming stations on the third deck. This appears improbable inasmuch as the bombs in this station were reported in reference (a) to have been 1,000 pounds. Bombs of this size and type are not particularly sensitive to shock. Further, when bombs of this size do detonate in confined spaces next to the shell the shell itself would unquestionably be extensively ruptured by blast and fragments. No damage to the hull above the waterline in the vicinity of the bomb arming stations was reported and none is evident in any of the photographs.

37. The three explosions at 1449, 1505 and 1515 apparently resulted from the ignition of gasoline vapors released by the leaks in the gasoline delivery main at frame 63 on the second deck. Since the gasoline system was in use, the gasoline drained into this area and was ignited by the fires started from the torpedo hits and the first interior explosion discussed above. It was reported in reference (a) that the gasoline main "shattered" as a result of the shock imparted by the torpedo detonations. This is highly improbable as the gasoline mains were made of copper with silver soldered joints and are highly resistant to shock. It appears more probable that the blast of the first interior explosion distorted the supporting structure sufficiently to break the main at fittings or that fragments pierced the pipe permitting gasoline to spill into the second and third deck spaces at frame 63.

38. The rupture of fuel oil and gasoline tanks by the torpedo detonations undoubtedly resulted in the spreading of inflammable liquids by flooding throughout the damaged area below the fourth deck as well as on the surface of the water

abreast the hits. Ignition of the vapor could have been caused by a spark from any one of a number of sources or by hot fragments or shorts in energized electrical circuits. The oil and gasoline on the surface of the water were then ignited in turn.

39. The fire in the hangar in all probability resulted from ruptured gasoline tanks in airplanes. The lashings of the planes triced to the overhead carried away and the planes crashed onto those below on the hangar deck. Under the circumstances the gasoline tanks of planes could scarcely have escaped damage, and probably gasoline ran out on the deck. Its ignition would almost inevitably have resulted under the circumstances because of loose debris and sparks from the fires below. The fire in the hangar rapidly spread among the aircraft. Gasoline from the fuel tanks of previously undamaged aircraft undoubtedly augmented the fire. Depth bombs on the armed aircraft must have added to the conflagration although no detonation of any of these seems to have occurred. The cases of the depth bombs undoubtedly split open from heat and the charges burned violently.

40. The fire in the hangar spread up into the forward starboard gun group where a short time later the 5-inch projectiles began to detonate. A survivor reported an explosion in this gun group. It is possible that powder charges in one or more ready-service boxes caught fire and built up pressure within the box so as to burst it in a manner resembling an explosion. On the other hand the box would probably not withstand much internal pressure. Hence it seems more probable that the explosion reported was a detonation of one of the 5-inch projectiles. Other cases of war damage have demonstrated that 5-inch projectiles will heat up and detonate in severe fires, after a period of roughly fifteen to twenty minutes. The fire in this area was sufficiently intense to detonate the 5-inch projectiles whether nearby powder burned or not, but it is probable that the powder also burned without necessarily producing any explosions.

41. Mention was made in reference (a) that the wood on the flight deck was on fire. This is another of the few instances reported of a wood deck catching fire. On ERIE* the wood main deck was consumed by fire. Instances where wood decks have been charred are more usual. The degree of burning depends, of course, on the intensity of the fire and the length of time involved.

42. In conjunction with the fires and explosions, it was reported by the Medical Officer that 99% of the injuries were burns of varying degrees of severity, mostly about the face, upper extremities and trunk.

E. Damage to the Fire Main System

43. The Engineering Officer reported in reference (a) that, "breakage of the damage control main and fire mains prevented delivery of water to the vicinity of the fire". The reports of other survivors indicate, however, beyond a doubt that a

---

*BuShips War Damage Report No. 31.

large portion of the fire main system aft of frame 100 was intact. This is evidenced by the fact that at least two hose lines were used and these were connected to plugs at frames 98 and 105 on the hangar deck, although the pressure at these hoses was very low. It appears probable that the starboard section of the fire main loop forward of the forward engine room was cut by the detonations of the torpedoes and that this section was not isolated, although out-out valves for the port and starboard sections of the forward loop were located in the forward engine room at frame 76. With the exception of the turbo generator, the dynamo condenser and the main distribution board, the machinery in the forward engine room remained operable and in an undamaged condition, and the engine room watch continued at their stations until WASP was abandoned. It also appears that Nos. 3, 4, and 5 water curtains on the hangar deck were not placed in operation because the steam-driven damage control pump in the after engine room was not placed on the supply risers to the curtains, although this pump was warmed up and in a standby condition at the time WASP was torpedoed. It is possible that the damage control pump was utilized for reducing the 11 degree list to 4 degrees as the Commanding Officer reported that tanks C-708-F and C-710-F were filled from the damage control main. The Commanding Officer's report also indicated that some water was delivered to No. 3 water curtain, at frame 104, although the water supply lasted for only a few seconds. This indicates that the riser for this water curtain was intact. The source of water was probably the fire main. No mention was made by survivors of efforts to use the hangar sprinkling system which is supplied by the damage control system. This system consists of sprinklers in the hangar overhead located between water curtains and fed by the damage control main or the fire main. It appears probable that all water curtains and sprinklers, and piping thereto, were intact from No. 2 water curtain at frame 73 to the aft end of the hangar - at least until the two vapor explosions "B" and "C" occurred about twenty minutes after the torpedo hits. If the riser at frame 76 was broken by these vapor explosions, the water curtains, sprinklers and piping aft of and including No. 3 water curtain at frame 99 should have remained intact thereafter.

44. WASP had an excellent fire main system. There was one 500 g.p.m. steam-driven fire and flushing pump in each engine room, and a total of five steam-driven 200 g.p.m. fire and bilge pumps located one in each engine room and one in each of the three boiler operating stations. The total pumping capacity was thus 2,000 g.p.m. at a pressure of 150 pounds per sq.in.. The fire main proper was subdivided into three complete loops below the first platform. One loop, which undoubtedly was cut by the torpedo detonations, extended from forward of bulkhead 76 (the forward bulkhead of the forward engine room) to the forward magazines. The second loop made a complete circuit of the main engineering spaces, and the third loop extended aft from bulkhead 110-1/2 (the after bulkhead of the after engine room) to the after magazines. The various loops were provided with adequate cross-connections and cut-out valves so that a damaged section of the system could be isolated readily. WASP also was provided with two 10,000 g.p.m. damage control pumps capable of delivering water at a head of 100 feet. One of these was steam-driven and

located in the after engine room and the other was motor-driven and located in the forward engine room. The damage control pumps could be utilized for a wide variety of purposes. Thus, they could be connected to any of the following systems as desired: the fire main, the fuel-oil filling and transfer system, the drainage system and the sprinkling system and water curtains in the hangar. There were two sources of supply to the sprinkling system and water curtains: one, as noted above, from the damage control system, and the other from the fire main system. Either source could be used. As far as could be determined, it appears that all of these facilities were available with the exception of the forward loop of the fire main and the motor-driven damage control pump which may have been temporarily inoperative because of the casualty to the forward switchboard. In other words, pumping capacity of 2,000 g.p.m., sufficient for six large (2-1/2-inch diameter) hoses, was available to the two after loops of the fire main, and 10,000 g.p.m. from the after damage control pump could have been used for the water curtains and sprinkling system. The fact that the forward loop was broken would reduce the pressure on the balance of the fire main system to a negligible value if the damaged loop were not cut out.

45. It is noteworthy that in the case of SAVANNAH, which recently received very severe damage from a large A.P. bomb, the forward fire main was broken by the direct effects of the bomb explosion, but this broken portion was isolated by closing cut-off valves. Although the main sources of electrical power failed, emergency means were used to get full pressure from several fire hoses at the scene of the fire within about six minutes after the bomb detonation. All fires were extinguished in a reasonably short time. It should be noted, however, that these were not gasoline fires.

46. A survivor from WASP reported that, "the fire main was ruptured in the trunk to central and not in the torpedoed spaces". He stated in connection with this that, "the reason for this fracture, I feel, was due to cast iron piping". This Bureau has but one authentic report in its files of war damage which tends to substantiate in even the slightest degree the statements quoted above. The one specific case of fracture occurred on STERETT*. In the case of STERETT a branch connection welded in the fire main piping and located in No. 1 fireroom was fractured by the shock of the detonation of an enemy projectile. The failure was in the weld and possibly may have been due to defective weld metal. It is noted that this damage was caused by shock from a projectile and not from a torpedo. On the other hand, there are reports available which indicate that numerous combatant ships which have been torpedoed did not suffer fracture of the fire main as a result of shock. To name a few, SARATOGA, CHESTER, PENSACOLA, PORTLAND, O'BRIEN and NORTH CAROLINA all were torpedoed and, in all of these cases, the fire main piping remained substantially intact except in the area subjected to direct attack by fragments or the blast effect from very close detonations. It is also pointed out that the fire main on WASP was of tough ductile material which

- -

*BuShips War Damage Report No. 33

cannot be shattered like a brittle material such as glass or cast iron. Fire main piping on WASP was galvanized wrought iron It is thus exceedingly difficult to reconcile the quoted statements with the known characteristics of the system in question, and the known performance of similar systems on other vessels.

F. Stowage of Aircraft in the Overhead Bays of the Hangar

47. It has been noted in paragraphs 4, 14 and 39 that the tricing arrangements for four planes triced to the overhead in the forward portion of the hangar carried away because of shock and permitted the planes to crash onto other aircraft parked on the deck. The plane gasoline tanks could not have escaped damage and gasoline must have spread into the hangar. This gasoline was unquestionably the origin of the fire in the hangar.

48. War experience on other carriers subjected to bombing attack has demonstrated that fires among triced aircraft are not only apt to occur but that such fires are very difficult to control and extinguish because of the overhead location. Fires among triced aircraft occurred on both ENTERPRISE and HORNET, and in both cases were difficult to extinguish.

49. The dangers of overhead stowage for planes have been recognized since the early days of the present war, and on 21 March, 1942 the Secretary of the Navy approved the design characteristics for the CVB41 class which stated, "reserve airplane stowage to be provided, if practicable, on the same level as and adjacent to the ends of the hangar to eliminate plane stowage overhead".

50. Overhead plane stowage has been omitted on some of the vessels of the CV9 class because of the strengthening of the flight deck required to permit operation of the latest types of heavy planes. Strengthening involved the addition of girders under the flight deck which prohibited the installation of tricing arrangements. The removal of tricing arrangements from all other carriers in service or under construction is now under serious consideration.

G. Improvements in Gasoline Systems of Aircraft Carriers.

(Plates III and IV)

51. As noted in paragraph 7 of the summary, an exhaustive study of gasoline stowage and gasoline systems aboard carriers has been made. Many improvements have been developed and are being made on carriers now under construction. Where practicable, changes will be made in existing carriers.

52. For the purpose of discussion, these improvements have been subdivided into four phases:

(a) The protection of gasoline stowage against underwater attack, shells, bombs, splinters, etc.,

(b) The means of limiting the spread of gasoline and vapors once the given protection is penetrated,

(c) The equipment and devices for fighting and controlling such fires if and when they occur,

(d) The thorough training of operating personnel in order that the handling of gasoline will be accomplished with maximum safety and that the proper operation of the facilities provided will be insured.

53. Improvements in protection:

(a) In large carriers, gasoline stowage tanks will be located within the main armored protection and the fully developed torpedo protection system. This will be done in the later vessels of the CV9 and CVB41 classes. Underwater protection for the light carriers CVL22 and CVL48 classes and for the escort carriers CVE6 and CVE55 classes is necessarily limited by the basic characteristics of the types. In the CVL22 class a blister was added for stability and some increase in underwater protection thus resulted; but, in none of these classes of light and escort carriers is the underwater hull sufficient to resist penetration by contact torpedo explosions. Hulls of the CVE26 and CVE105 classes are similar to those of modern oil tankers and inherently possess a fair measure of resistance to fatal torpedo damage because of the rather heavy hull construction and large depth of liquid carried next to the shell. Comparison of plate III with plate IV will give an approximate idea of the increased protection afforded in new CVs beyond that given WASP.

(b) The gasoline stowage tanks will be subdivided into three tanks similar to the arrangement shown on plate III. The provision of draw-off tanks permits increasing the total effective stowage capacity by several percent without increasing the gasoline stowage space. In addition, as gasoline is used the protection against underwater explosions becomes more effective. This "saddle tank" arrangement has been designed for the forward tanks of the CVB41 class and will be fitted in the stowage of the later vessels of the CV9 class and the CVE105 class. The after gasoline stowage in the CVB41 class is not of size and shape to permit the use of the "saddle tank" to advantage. This stowage, however, is well in from the sides of the ship and inboard of the outboard shaft alleys. A modified "saddle tank" arrangement will be installed in the new CVL48 Class.

(c) The inboard gasoline piping mains will be made as short as possible and will be in an STS protected trunk from the point where they penetrate the armored deck to the point where they emerge from the ship through the shell plating. This installation will be followed for the CVB41 class and for the later vessels of the CV9 and CVL classes. These trunks are being installed on the CVE105 class but are not of STS.

54. Improvements to prevent the spread and ignition of gasoline vapor:

(a) Improvements in subdivision by the omission or elimination of doors and other openings through vertical boundaries on the lower decks will tend to prevent the spread of explosive vapors after battle damage.

(b) The installation of ventilation piping to voids surrounding gasoline stowages is being discontinued on new carriers. It will be removed on existing carriers as soon as practicable. This piping has been a potent means for spreading gasoline vapors through the ship after damage.

(c) Drainage piping for gasoline stowage tanks and the surrounding voids will be modified in order to minimize piping penetrations of the boundaries. This piping has also been a means for spreading gasoline vapors through the ship after damage.

(d) Each void will be fitted with its own permanently installed eductor, with the supply and discharge piping terminating at hose connections within watertight boxes located in the pump room. These boxes afford additional watertight protection where the void boundaries are penetrated. The eductors are actuated by a fire hose from a nearby fire plug and where possible will discharge through a hose connected to the salt water overflow line of the gasoline system, otherwise discharge will be directly overboard (see plate III). The gasoline tanks are not fitted for drainage and portable eductors are to be used whenever drainage is necessary.

(e) Voids surrounding gasoline tanks will be charged with an inert gas, i.e., one that will not support combustion. This will reduce the hazard of the presence of gasoline vapors and possible leaks in adjacent gasoline tanks. This protection will be effective, however, only so long as the boundaries of the voids are gas-tight. Once these boundaries are ruptured or damaged so that the inert gas and vapor will flow into an adjoining compartment, it is known that sufficient oxygen will be present to form inflammable concentrations.

(f) Electrical equipment which would cause ignition of gasoline vapor will not be installed in compartment adjacent to these voids because gasoline vapors may be present in the voids, which vapors are kept non-explosive by the inert gas only as long as the proper mixture obtains. If the boundaries of the voids are ruptured, the inert gases will be diluted and explosive mixtures may easily develop. This provision eliminates the possibility of ignition from this source in case of damage to void boundaries. Where electrical fittings must be installed in these adjacent compartments, they will be required to have explosion-proof characteristics.

(g) Where these "adjacent" spaces require ventilation, they will be ventilated by independent systems which will lead directly to or from the weather deck and will not serve other spaces not "adjacent". Two or more of these "adjacent" spaces may be served from the same system when they are located between the same main transverse bulkheads.

(h) The gasoline trunk and pump room will be ventilated by an independent system which leads directly to or from the weather deck. This system may also serve the pump motor room, the ducts being guarded by flame-proof screens.

(i) The gasoline supply main within the ship will be a double-walled pipe. The inner pipe will carry the gasoline and the space between the inner and outer pipe will be filled with a fire extinguishing agent under pressure.

(j) Means are being provided for filling the mains with inert gas under pressure when the gasoline has been drained. This is primarily a safety precaution to insure detection of leaks in the system rather than a means of increasing resistance to damage. Flushing with water is of questionable value from the standpoint of fire protection. Beginning with the CV9 class the mains can be blown empty by the inert gas fed into them at 20 pounds pressure. Actually on test, mains on the CV9 have been emptied by this means in 7 minutes.

(k) Inert gas pressure gages are being supplied in order to permit ready detection of leaks by observation without the necessity of pumping gasoline through the lines.

(l) Two or more electrically-driven gasoline pumps, with the motors located in separate watertight compartments, will be used in lieu of single large pumps. This will give a more flexible pumping system with the obvious advantage of reducing the rate of delivery of gasoline to the system when only a portion of the system is in use and hence reducing the rate of discharge in case a line in use is broken. This will be done on CVB41 class.

(m) The piping and valves within the gasoline pump rooms will be arranged to permit the use of the pumps to accelerate the drainage of the mains.

(n) The salt water compensating lines between the pump rooms and the tanks are being reduced to a minimum in order to limit the penetration of the tank boundaries.

55. Improvements in firefighting:

(a) A complete discussion of the latest developments in firefighting equipment and technique are contained in the Bureau's letter, file No. C-S93-(1)(3688)(8688), C-EN28/A2-11 of 25 February, 1943 which has been distributed to all vessels.

(b) The Firefighting Manual, issued by the Bureau in 1943, contains the latest instructions regarding the shipboard use of the Navy's standard firefighting equipment and a complete description of the technique for fighting fire. Much of the equipment as described in this publication is either newly designed or redesigned to embody the latest improvements.

(c) Firefighting schools - see paragraph 56(c).

56. Training of personnel:

(a) The Bureau is in the process of preparing well illustrated gasoline system operating manuals for use on carriers.

(b) The establishment of schools for special training in gasoline handling has been approved. Complete working models of ship gasoline systems and all associated apparatus, such as inert gas systems, are nearing completion for use in these schools.

(c) The firefighting schools which have been well established offer the opportunity at all major yards and bases for thorough training of key personnel, both officer and enlisted, in combatting shipboard fires using the most modern equipment and the latest technique.

H. Design Characteristics of WASP

57. WASP was designed during 1934 and 1935. The displacement of WASP was limited to 14,700 tons by the restrictions on total tonnage of aircraft carriers imposed by the Limitation of Naval Armament treaties then in force. It is not possible to provide in a vessel of this size a fully-developed system of underwater protection. In an effort to overcome this handicap, WASP was subdivided into relatively small compartments and a liquid layer inboard of the shell was provided. Although the primary purpose of the liquid layer is to reduce the angle of list following underwater attack, it has been found that the liquid layer will also reduce the velocity of fragments from underwater explosions to the point where there is little or no danger of initiating detonations among explosives carried within the magazines. In addition, a liquid layer will reduce the extent of structural damage to the shell. Apparently, the liquid layer on WASP in way of the torpedo detonations was full, either of fuel oil or water, at the time of the attack. The value of such a layer was well demonstrated in this case inasmuch as the angle of list assumed was not excessive; and furthermore, the detonation of the torpedo at frame 62 did not cause detonation of the bombs in the bomb magazines, although the latter were directly in way of the torpedo.

58. The new CVL's, converted in 1941 and 1942 from cruisers of the CL55 class, are somewhat smaller than WASP. Their underwater layout is much the same as for WASP except that a blister was added. This will give a deeper liquid layer. However, in the event of torpedo attack, it can be expected that rupture of the gasoline tanks will occur if the torpedo strikes in way of them. The improvements in gasoline stowage and handling discussed in part F, however, will result in much better resistance to gasoline vapor fires and explosion.

I. Conclusions

59. WASP, in spite of her small size, would have survived the damage to the hull had there been no fires. Watertight subdivision, stability characteristics and reserve buoyancy were all adequate to absorb the damage from two torpedoes, located as these were, without fatal consequences. It is noted that for some time after the attack WASP was still in a condition to proceed from the scene of action under her own power

if the fires had been controlled. After she was abandoned she was struck by three additional torpedoes. Even after this, she did not sink immediately, although by then she was completely gutted by fire. Considering the durability of WASP, it was unfortunate that fire caused her loss.

(a) Full use was not made of the firefighting facilities available for various reasons as discussed in this report.

(b) It is problematical whether the intense gasoline fires could have been controlled even with full use of available facilities.

60. With respect to gasoline stowage and handling, design improvements in later carriers of larger size have been directed toward:

(a) Improving protection of gasoline tanks and systems with the objectives of reducing the probability of fires following damage in battle.

(b) The improvement of firefighting facilities to permit better control of gasoline fires if they do occur.

61. It should be remembered, however, that gasoline is an extremely hazardous liquid and will continue to be so in spite of all precautions that can be taken in design. The utmost vigilance on the part of operating personnel will be required even in our newest ships during normal operations and particularly so after damage has been incurred.

www.ingramcontent.com/pod-product-compliance
Ingram Content Group UK Ltd.
Pitfield, Milton Keynes, MK11 3LW, UK
UKHW052226270726
14059UKWH00003B/145